Sequential Approximate Multiobjective Optimization Using Computational Intelligence

Series Editor:

Johannes Jahn
University of Erlangen-Nürnberg
Department of Mathematics
Martensstr. 3
81058 Erlangen
Germany
jahn@am.uni-erlangen.de

Vector Optimization

The series in Vector Optimization contains publications in various fields of optimization with vector-valued objective functions, such as multiobjective optimization, multi criteria decision making, set optimization, vector-valued game theory and border areas to financial mathematics, biosystems, semidefinite programming and multiobjective control theory. Studies of continuous, discrete, combinatorial and stochastic multiobjective models in interesting fields of operations research are also included. The series covers mathematical theory, methods and applications in economics and engineering. These publications being written in English are primarily monographs and multiple author works containing current advances in these fields.

Hirotaka Nakayama · Yeboon Yun · Min Yoon

Sequential Approximate Multiobjective Optimization Using Computational Intelligence

Professor Hirotaka Nakayama
Dept. of Intelligence and Informatics
Konan University
8-9-1 Okamoto, Higashinada
Kobe 658-8501
Japan

Assistant Professor Min Yoon
Division of Mathematical Science
Pukyong National University
599 Daeyeon 3-dong, Nam-gu
Busan 608-737
Korea

Associate Professor Yeboon Yun
Dept. of Reliability-based Information
   System Engineering
Faculty of Engineering
Kagawa University
2217-20 Hayashi-cho
Takamatsu
Kagawa 761-0396
Japan

ISBN 978-3-540-88909-0   e-ISBN 978-3-540-88910-6

DOI 10.1007/978-3-540-88910-6

Springer Series in Vector Optimization ISSN 1867-8971

Library of Congress Control Number: 2008943999

© Springer-Verlag Berlin Heidelberg 2009

This work is subject to copyright. All rights are reserved, whether the whole or part of the material is concerned, specifically the rights of translation, reprinting, reuse of illustrations, recitation, broadcasting, reproduction on microfilm or in any other way, and storage in data banks. Duplication of this publication or parts thereof is permitted only under the provisions of the German Copyright Law of September 9, 1965, in its current version, and permission for use must always be obtained from Springer. Violations are liable to prosecution under the German Copyright Law.

The use of general descriptive names, registered names, trademarks, etc. in this publication does not imply, even in the absence of a specific statement, that such names are exempt from the relevant protective laws and regulations and therefore free for general use.

Cover design: WMXDesign GmbH, Heidelberg

Printed on acid-free paper

springer.com

Preface

Many kinds of practical problems such as engineering design, industrial management and financial investment have multiple objectives conflicting with each other. Those problems can be formulated as multiobjective optimization. In multiobjective optimization, there does not necessarily a unique solution which minimizes (or maximizes) all objective functions. We usually face to the situation in which if we want to improve some of objectives, we have to give up other objectives. Finally, we pay much attention on how much to improve some of objectives and instead how much to give up others. This is called "trade-off." Note that making trade-off is a problem of value judgment of decision makers. One of main themes of multiobjective optimization is how to incorporate value judgment of decision makers into decision support systems. There are two major issues in value judgment (1) multiplicity of value judgment and (2) dynamics of value judgment. The multiplicity of value judgment is treated as trade-off analysis in multiobjective optimization. On the other hand, dynamics of value judgment is difficult to treat. However, it is natural that decision makers change their value judgment even in decision making process, because they obtain new information during the process. Therefore, decision support systems are to be robust against the change of value judgment of decision makers. To this aim, interactive programming methods which search a solution while eliciting partial information on value judgment of decision makers have been developed. Those methods are required to perform flexibly for decision makers' attitude. At early 1980s, many interactive programming methods for solving multiobjective optimization have been developed. Above all, the aspiration level approach to multiobjective programming problems has been widely recognized to be effective in many practical fields.

Another major issue is that in many practical problems, in particular in engineering design, the function form of criteria is not given explicitly in terms of design variables. Given the value of design variables, under this circumstance, the value of objective functions is obtained by real/computational experiments such as structural analysis, fluid mechanic analysis, thermodynamic analysis, and so on. Usually, these experiments are time consuming and expensive. One of recent trends in optimization is how to treat these expensive criteria. In order to make the number of these experiments as few as possible, optimization is performed in parallel with predicting the form of objective functions. This is called sequential approximate optimization with metamodeling. It has been observed that techniques of computational intelligence can be effectively applied for this purpose. Moreover, techniques of multiobjective optimization themselves can also be applied to develop effective methods in computational intelligence. For example, the authors developed several kinds of support vector machines using multiobjective optimization and goal programming.

Recently, researches of generating Pareto frontier are actively made. It is useful to visualize Pareto frontier, because decision makers can make tradeoff analysis very easily on the shown figures of Pareto frontier. However, it is difficult to generate Pareto frontier in cases with more than two or three objectives. At this event, a method combining aspiration level approach and sequential approximate optimization using computational intelligence was proposed and recognized to be effective in many practical problems.

This book describes those sophisticated methods for multiobjective optimization using computational intelligence along with real applications. This topic seems quite new. No book on this topic has been seen to our knowledge in spite of its importance. The book is self-contained and comprehensive. The potential readers are researchers, practitioners in industries and students of graduate course and high grade of undergraduate course.

We hope that the readers of both theoretical researchers and practitioners can learn through this book several methodologies on a new trend of multiobjective optimization which is very important and applicable in many practical fields.

Japan and Korea
November 2008

Hirotaka Nakayama
Yeboon Yun
Min Yoon

Contents

List of Tables xi

List of Figures xiii

1 Basic Concepts of Multiobjective Optimization 1
  1.1 Mathematical Foundations 1
  1.2 Preference Order and Domination Set 4
  1.3 Scalarization 5
    1.3.1 Linearly Weighted Sum 7
    1.3.2 Tchebyshev Scalarization Function 8
    1.3.3 Augmented Tchebyshev Scalarization Function 8
    1.3.4 Constraint Transformation Method 10
  1.4 Scalarization and Trade-off Analysis 11

**2 Interactive Programming Methods for Multiobjective
   Optimization** 17
  2.1 Goal Programming 17
  2.2 Why is the Weighting Method Ineffective? 20
  2.3 Satisficing Trade-off Method 22
    2.3.1 On the Operation P 22
    2.3.2 On the Operation T 24
    2.3.3 Automatic Trade-off 25
    2.3.4 Exact Trade-off 29
    2.3.5 Interchange Between Objectives and Constraints 29
    2.3.6 Remarks on Trade-off for Objective Functions
          with 0-Sensitivity 30
    2.3.7 Relationship to Fuzzy Mathematical Programming ... 32
  2.4 Applications 34
    2.4.1 Feed Formulation for Live Stock 34
    2.4.2 Erection Management of Cable-Stayed Bridge 35
    2.4.3 An Interactive Support System for Bond Trading 37
  2.5 Some Remarks on Applications 42

viii   Contents

3 Generation of Pareto Frontier by Genetic Algorithms 45
   3.1 Evolutionary Multiobjective Optimization 45
      3.1.1 Vector Evaluated Genetic Algorithm 46
      3.1.2 Multiobjective Genetic Algorithm 47
      3.1.3 Elitist Nondominated Sorting Genetic Algorithm
            (NSGA-II) 51
      3.1.4 Strength Pareto Evolutionary Algorithm (SPEA2) 53
   3.2 Fitness Evaluation Using DEA 56
      3.2.1 Quantitative Evaluation of Fitness.................. 56
      3.2.2 Data Envelopment Analysis........................ 57
      3.2.3 DEA Models...................................... 59
      3.2.4 DEA Method 63
   3.3 Fitness Evaluation Using GDEA 64
      3.3.1 GDEA Model 64
      3.3.2 GDEA Method 66
   3.4 Comparisons of Several Fitness Evaluations 67

**4 Multiobjective Optimization and Computational
Intelligence** 73
   4.1 Machine Learning 73
      4.1.1 Learning and Generalization 75
      4.1.2 The Least Square Method 77
   4.2 Radial Basis Function Networks 79
   4.3 Support Vector Machines for Pattern Classification 83
      4.3.1 Hard Margin SVM 84
      4.3.2 MOP/GP Approaches to Pattern Classification 86
      4.3.3 Soft Margin SVM 88
      4.3.4 ν-SVM.. 90
      4.3.5 Extensions of SVM by MOP/GP 90
      4.3.6 Comparison of Experimental Results................ 96
   4.4 Support Vector Machines for Regression 98
   4.5 Combining Predetermined Model and SVR/RBFN.......... 110

5 Sequential Approximate Optimization 113
   5.1 Metamodels 113
   5.2 Optimal Design of Experiments 115
   5.3 Distance-Based Criteria for Optimal Design............... 120
   5.4 Incremental Design of Experiments 122
   5.5 Kriging and Efficient Global Optimization 126
      5.5.1 Kriging and Prediction............................ 127
      5.5.2 Efficient Global Optimization 133
   5.6 Distance-Based Local and Global Information 141

Contents ix

6 Combining Aspiration Level Approach and SAMO 151
  6.1 Sequential Approximate Multiobjective Optimization
      Using Satisficing Trade-off Method 152
  6.2 MCDM with Aspiration Level Method and GDEA 159
  6.3 Discussions 167

7 Engineering Applications 169
  7.1 Reinforcement of Cable-Stayed Bridges 169
      7.1.1 Dynamic Characteristics of Cable-Stayed Bridge...... 170
      7.1.2 Discussions 174
  7.2 Multiobjective Startup Scheduling of Power Plants 176
      7.2.1 Startup Scheduling for Combined Cycle Power Plant .. 176
      7.2.2 Sequential Approximate Multiobjective Optimization
           for Optimal Startup Scheduling 178
      7.2.3 Application Results.............................. 179
      7.2.4 Discussions 183

References 185

Index 193

List of Tables

3.1	Fitness assignment of MOGA	50
3.2	Case of single input and single output	58
3.3	DEA efficiency of Example 3.1	59
4.1	Classification rate by GP	97
4.2	Classification rate by SVM_{hard}	97
4.3	Classification rate by SVM_{soft}	98
4.4	Classification rate by ν-SVM	99
4.5	Classification rate by SVM_{total}	100
4.6	Classification rate by μ-SVM	101
4.7	Classification rate by μ-ν-SVM	102
4.8	Comparison of the results (unit: %)	110
5.1	Results by distance-based local and global information in Example 5.5	144
5.2	Simulation results of example (5.29) with $C_x^0 = 50$ and $C_f^0 = 30$	148
5.3	Comparison among several methods	149
6.1	Result by SQP using a quasi-Newton method for real objective functions	157
6.2	Result by SAMO with 100 function evaluations	158
7.1	Result for Case 1	174
7.2	Result for Case 2	175
7.3	Approximation error of the RBFN models (unit: %)	181
7.4	Results of the objective functions	183

List of Figures

1.1   Vector inequality 2
1.2   Pareto solutions (*bold curve*) in the objective space 3
1.3   Weak Pareto solutions in the objective space 3
1.4   Order preserving scalarization function 7
1.5   Pareto solutions in nonconvex parts cannot be obtained
      by any linearly weighted sum scalarization function 7
1.6   Tchebyshev scalarization function 8
1.7   Augmented Tchebyshev scalarization function 9
1.8   Constraint transformation method 11

2.1   Difference between solutions associated with two kinds
      of weight 24
2.2   Automatic trade-off (case 1) 27
2.3   Automatic trade-off (case 2) 28
2.4   Automatic trade-off (case 3) 29
2.5   Exact trade-off 30
2.6   Case of automatic trade-off with 0-sensitivity 31
2.7   Membership function for f_1 in (F) 32
2.8   Phase of an erection management system 36

3.1   Flowchart of GA 46
3.2   Schematic of VEGA 47
3.3   Ranking methods: each number in *parentheses* represents
      the rank of each individual and the *curve* represents
      the true Pareto frontier 48
3.4   Sharing function and niche count: for example,
      $nc_4 = \sum_{j=1}^{7} Sh_{4j} = 0 + 0 + 0.4 + 1 + 0.7 + 0.1 + 0 = 2.2$,
      where $Sh(d_{4j})$ is denoted by Sh_{4j} for simplicity 49
3.5   Data of Table 3.1 50
3.6   Schematic of NSGA-II 52

xiv | List of Figures

3.7   Crowding distance in NSGA-II: for example, the crowding distance of x_j in the objective function space is given by $d_{j1} + d_{j2}$ 52

3.8   Raw fitness assignment in SPEA2: A is dominated by B and C, where B dominates two individuals, while C three individuals. Thus the raw fitness for A is assigned by the sum of $R(A) = S(B) + S(C) = 5$ 54

3.9   Archive truncation method: for the nearest two points A and B, they have the next nearest distance, d_A and d_B. In this case, B is deleted since $d_A > d_B$ 55

3.10   A problem of rank-based methods: the dotted curve is a Pareto frontier, and the rank does not necessarily reflect the distance from the Pareto frontier 57

3.11   Efficient frontier by DEA (Example 3.1): the *solid line* is the DEA-efficient frontier 59

3.12   Efficient frontier by BCC model 62

3.13   Efficient frontier by FDH model 62

3.14   DEA method 63

3.15   GDEA-efficient frontiers by the parameter α in GDEA model.... 65

3.16   Geometric interpretation of α in (GDEA$_{fit}$): *points* are individuals, the *solid line* is GDEA-efficient frontier, and the *dotted line* is an approximate Pareto frontier 67

3.17   Results for Example 3.2..................................... 69

3.18   Results for Example 3.3..................................... 70

3.19   Results for Example 3.4..................................... 71

4.1   Radial basis function networks............................. 79

4.2   Gauss functions... 80

4.3   Classification by radial basis function networks 81

4.4   Regression by radial basis function networks: the *solid line* depicts the predicted function, and the *dotted line* does the true function 82

4.5   Mapping of data from an original input space to a feature space.. 84

4.6   Geometric interpretation of margin......................... 85

4.7   Classification by hard margin SVM......................... 86

4.8   Slack variable (exterior deviation) 88

4.9   Classification by soft margin SVM: the *solid line* represents the discriminant function and the *symbol* $*$ does support vectors 89

4.10   Classification by ν-SVM: the *solid line* represents the discriminant function and the *symbol* $*$ does support vectors 91

4.11   Classification by μ-ν-SVM: the *solid line* represents the discriminant function and the *symbol* $*$ does support vectors..... 95

List of Figures

4.12 ε-insensitive loss function 102
4.13 Support vector regression 103
4.14 Regression functions by C-SVR: the *dotted line* is the true function, the *thick line* a predicted function $y = f(x)$, and *solid lines* are $y = f(x) \pm \varepsilon$. The *symbol* $*$ shows support vectors 104
4.15 Regression functions by ν-SVR: the *dotted line* is the true function, the *thick line* a predicted function $y = f(x)$, and *solid lines* are $y = f(x) \pm \varepsilon$. The *symbol* $*$ shows support vectors 105
4.16 Regression functions by μ-SVR: the *dotted line* is the true function, the *thick line* a predicted function $y = f(x)$, and *solid lines* are $y = f(x) \pm \varepsilon$. The *symbol* $*$ shows support vectors 107
4.17 Regression functions by ν_ε-SVR: the *dotted line* is the true function, the *thick line* a predicted function $y = f(x)$, and *solid lines* are $y = f(x) \pm \varepsilon$. The *symbol* $*$ shows support vectors 109
4.18 Linear polynomial regression 111
4.19 Regression by μ-SVR 111
4.20 Result by combining linear polynomial model and μ-SVR with the same parameter of Fig. 4.19 112
4.21 Comparison of RMSE: r is a radius in Gauss kernel function 112

5.1 Confidence ellipsoid and several criteria for optimal design [48] .. 118
5.2 Max–min and min–max distance criteria for optimal design [63] .. 121
5.3 Additional point by distance-based optimality criteria 123
5.4 D-optimal designs for several models (no. of sample data $= 4$) 124
5.5 D-optimal designs for several models (no. of sample data $= 5$) 124
5.6 D-optimal designs for several models (no. of sample data $= 6$) 124
5.7 D-optimal designs for several models (no. of sample data $= 7$) 125
5.8 D-optimal designs for several models (no. of sample data $= 8$) 125
5.9 D-optimal designs for several models (no. of sample data $= 100$) .. 125
5.10 Distance-based designs for incremental experiments 126
5.11 Prediction by ordinary Kriging: the *dotted line* represents the true function, the *dashed line* does the predicted value $\hat{y}$ and the *solid line* does the variance $\hat{s}^2$ of prediction providing information of uncertainty 133
5.12 Expected improvement (**a**) and $\hat{s}$ (**b**): the *dotted line* represents the true function and the *dashed line* does the mean of prediction 135
5.13 Expected improvements for several cases 137
5.14 Objective function in Example 5.5 138
5.15 No. of data $= 5$ (first iteration) 139
5.16 No. of data $= 7$ 139
5.17 No. of data $= 21$ 140
5.18 No. of data $= 35$ 140
5.19 No. of data $= 49$ 140
5.20 No. of data $= 63$ (final iteration) 141

5.21	Convergence process of $\hat{f}^*$ and f_{min} ($\hat{f}^*$: optimal value for the predicted objective function, f_{min}: the current best value among sample data)	141
5.22	Geometric interpretation of two additional points	142
5.23	No. of data $= 5$ (initial iteration)	144
5.24	No. of data $= 7$	144
5.25	No. of data $= 21$	145
5.26	No. of data $= 41$	145
5.27	No. of data $= 61$ (final iteration)	145
5.28	Convergence process of $\hat{f}^*$ and f_{min} ($\hat{f}^*$: optimal value for the predicted objective function, f_{min}: the current best value among sample data)	146
5.29	Pressure vessel	146
5.30	Convergence process of example (5.29)	148
6.1	Result with three points	154
6.2	Result with 13 points	154
6.3	Real contours of Example 6.1	155
6.4	No. of data $= 10$ (initial iteration)	155
6.5	No. of data $= 40$ (after 15 additional learning): Case 1	156
6.6	Welded beam design problem	157
6.7	Approximate Pareto frontier by SAMO	158
6.8	Pareto frontier by EMO (100 gen. $\times$ 50 pop.)	159
6.9	No. of data $= 40$ (after ten additional learning): Case 2	160
6.10	Geometric interpretation of the problem (ALGDEA$_{fit}$)	161
6.11	Result for Example 6.4	162
6.12	Welded beam design	163
6.13	Representative alternatives to the aspiration level	165
6.14	Alternatives to decision maker's requests	166
6.15	Alternatives to decision maker's additional requests	167
7.1	Cable-stayed bridge	170
7.2	Cable with additional mass	171
7.3	Examples of mode shape	172
7.4	Influence on bending moment by additional mass (no ratio of additional mass is fixed)	172
7.5	Influence on bending moment by additional mass (No. 19 ratio of additional mass is fixed at 0.5)	172
7.6	Plant configuration	177
7.7	Startup schedule for a power plant	178
7.8	Validation of RBFN model	181
7.9	Aspiration levels and corresponding solutions	181
7.10	Optimal startup schedule obtained for each aspiration level	182

Chapter 1
Basic Concepts of Multiobjective Optimization

Basic important concepts of multiobjective optimization was originated from Edgeworth and Pareto from the late of nineteenth century to the beginning of twentieth century, while a mathematical development was made by Cantor at almost the same ages. Today, we usually refer a solution of multiobjective optimization to as a Pareto solution. Scalarization techniques for multiple objective functions are already presented in Edgeworth's book [36] more than one hundred years ago. Several mathematical properties from a viewpoint of mathematical programming was developed by Kuhn and Tucker [77] in the middle of twentieth century. This chapter describes some essential topics of mathematical foundations in multiobjective optimization which will be used in the following chapters. Mathematical theory of multiobjective optimization can be seen in [47, 61, 130], and historical remarks in [24, 145].

1.1 Mathematical Foundations

Multiobjective programming problems are formulated as follows:

$$\underset{\boldsymbol{x}}{\text{minimize}} \qquad \boldsymbol{f}(\boldsymbol{x}) := (f_1(\boldsymbol{x}), \dots, f_r(\boldsymbol{x}))^T \qquad \text{(MOP)}$$
$$\text{subject to} \qquad \boldsymbol{x} \in X \subset \mathbb{R}^n.$$

The constraint set X may be given by

$$g_j(\boldsymbol{x}) \leqq 0, \quad j = 1, \dots, m,$$

and/or a subset of $\mathbb{R}^n$ itself.

H. Nakayama et al., *Sequential Approximate Multiobjective Optimization*
Using Computational Intelligence, Vector Optimization,
DOI 10.1007/978-3-540-88910-6_1, © 2009 Springer-Verlag Berlin Heidelberg

Throughout this book, we shall use the following vector inequalities:

Definition 1.1 (Vector Inequality). For any $y^1, y^2 \in \mathbb{R}^r$ (Fig. 1.1),

$$y^1 < y^2 \iff y_i^1 < y_i^2, \quad \forall\, i = 1, \ldots, r,$$
$$y^1 \leqq y^2 \iff y_i^1 \leqq y_i^2, \quad \forall\, i = 1, \ldots, r,$$
$$y^1 \leq y^2 \iff y^1 \leqq y^2,\ y^1 \neq y^2.$$

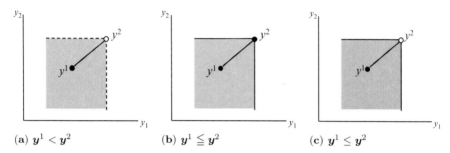

(a) $y^1 < y^2$ (b) $y^1 \leqq y^2$ (c) $y^1 \leq y^2$

Fig. 1.1 Vector inequality

Remark 1.1 (Positive (Negative) Orthant). Let the zero vector of any vector space be denoted by $\mathbf{0}$. A subset $\mathbb{R}^r_+ := \{\, y \in \mathbb{R}^r \mid y \geqq \mathbf{0}\,\}$ is called the *positive orthant* of $\mathbb{R}^r$, and $\mathbb{R}^r_- := \{\, y \in \mathbb{R}^r \mid y \leqq \mathbf{0}\,\}$ the *negative orthant* of $\mathbb{R}^r$. Similarly, we define the interior of positive orthant by $\operatorname{int} \mathbb{R}^r_+ := \{\, y \in \mathbb{R}^r \mid y > \mathbf{0}\,\}$, and that of the negative orthant $\operatorname{int} \mathbb{R}^r_- := \{\, y \in \mathbb{R}^r \mid y < \mathbf{0}\,\}$. Then

$$y^1 < y^2 \iff y^1 - y^2 \in \operatorname{int} \mathbb{R}^r_-,$$
$$y^1 \leqq y^2 \iff y^1 - y^2 \in \mathbb{R}^r_-,$$
$$y^1 \leq y^2 \iff y^1 - y^2 \in \mathbb{R}^r_- \setminus \{\mathbf{0}\}.$$

Here, the symbol $\mathbb{R}^r_- \setminus \{\mathbf{0}\}$ stands for the set removing $\{\mathbf{0}\}$ from $\mathbb{R}^r_-$.

Definition 1.2 (Pareto Solution). If there is no $x \in X$ such that $f(x) \leq f(\hat{x})$, $\hat{x}$ is referred to as *Pareto optimal*, or simply to as a *Pareto solution*.[1]

---

[1] The same notion as Pareto solution was introduced by Edgeworth in his famous book [36] in 1881. Important ideas of scalarization and constraint transformation method to be described in detail later in our book are found there. Pareto presented the idea of Pareto solution in 1906 [115]. The readers may refer the history of those ideas, in particular, from a viewpoint of utility theory in Stigler [148]. The name of efficient solution was given by Koopmans [74], nondominated solution by von Neumann–Morgenstern [154], and noninferior solution by Zadeh [164].

1.1 Mathematical Foundations

Hereafter, it would be more convenient to consider the Pareto optimality in the objective space like Fig. 1.2 rather than in the space of decision variables $\boldsymbol{x}$. Then the fact that $\hat{\boldsymbol{x}}$ is Pareto optimal means that the set $\hat{\boldsymbol{y}} + \mathbb{R}^r_-$ and $\boldsymbol{f}(X)$ have no common point except for $\hat{\boldsymbol{y}}\ (= \boldsymbol{f}(\hat{\boldsymbol{x}}))$.

The set of Pareto solutions in the objective space is referred to as *Pareto frontier* or *efficient frontier*. Usually we select one final solution from Pareto solutions taking the total balance among objectives into account.

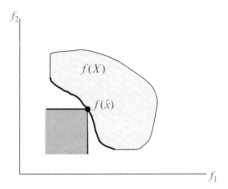

Fig. 1.2 Pareto solutions (*bold curve*) in the objective space

Remark 1.2. When there is no $\boldsymbol{x} \in X$ such that $\boldsymbol{f}(\boldsymbol{x}) < \boldsymbol{f}(\hat{\boldsymbol{x}})$, $\hat{\boldsymbol{x}}$ is called *weakly Pareto optimal*, or simply a *weak Pareto solution*. The fact that $\hat{\boldsymbol{x}}$ is a weak Pareto solution implies geometrically that the set $\hat{\boldsymbol{y}} + \text{int } \mathbb{R}^r_- \cup \{\mathbf{0}\}$ and $\boldsymbol{f}(X)$ have no common point except for $\hat{\boldsymbol{y}}(=\boldsymbol{f}(\hat{\boldsymbol{x}}))$. As is shown in Fig. 1.3, while fixing an objective at the level of the weak Pareto solution, we can improve another objective. Therefore, weak Pareto solutions are not welcome as solutions in actual decision making, but we often encounter cases in which only weak Pareto optimality is guaranteed theoretically.

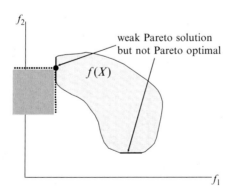

Fig. 1.3 Weak Pareto solutions in the objective space

In general, there may be many Pareto solutions. The final decision is made among them taking the total balance over all criteria into account. This is a problem of value judgment of decision maker (DM). The totally balancing over criteria is usually called *trade-off*. It should be noted that there are very many criteria, say, over one hundred in some practical problems such as erection management of cable stayed bridge, and camera lens design. Therefore, it is important to develop effective methods for helping DM to trade-off easily even in problems with very many criteria.

1.2 Preference Order and Domination Set

If a decision maker prefers y^1 to y^2, we denote this relation by $y^1 \succ y^2$. Clearly, the relation $\succ$ is a binary relation, and is called *preference relation*, or *preference order*. Furthermore, the relation $\sim$ defined by

$$y^1 \sim y^2 \iff not\ y^1 \succ y^2 \text{ and } not\ y^2 \succ y^1$$

is called an *indifference relation*.

For the preference $\succ$ and for $y \in \mathbb{R}^r$, the set

$$D^+(y) = \{d \in \mathbb{R}^r \mid y + d \succ y\}$$

is called the *domination set* for y, and

$$D^-(y) = \{d \in \mathbb{R}^r \mid y \succ y + d\}$$

is called the *dominated set* for y. In addition,

$$I(y) = \{d \in \mathbb{R}^r \mid y + d \sim y\}$$

is called the *indifference set* for y.

In our multiobjective optimization (MOP) in which each objective function is to be minimized, the domination set for Pareto order $\leq$ is given by

$$D^+(y) = \mathbb{R}^r_- \backslash \{0\}$$

for any $y \in \mathbb{R}^r$.

Remark 1.3. A nonempty closed convex cone $D \subset \mathbb{R}^r$ is called to be *acute*, if

$$D \cap (-D) = \{0\}.$$

Such an acute cone D induces a partial order. For example, we can define the relation $\leqq_D$ with $\mathrm{cl}\, D^+ = D$ and $\mathrm{cl}\, D^- = -D$ as follows:

$$y^1 \leqq_D y^2 \iff y^1 - y^2 \in D.$$

1.3 Scalarization 5

Similarly, $\leq_D$ with $D^+ = D \backslash \mathbf{0}$ and $D^- = -D \backslash \mathbf{0}$ is defined by

$$\mathbf{y}^1 \leq_D \mathbf{y}^2 \Longleftrightarrow \mathbf{y}^1 - \mathbf{y}^2 \in D \backslash \mathbf{0}.$$

Furthermore, $<_D$ with $D^+ = \operatorname{int} D$ and $D^- = -\operatorname{int} D$ is defined by

$$\mathbf{y}^1 <_D \mathbf{y}^2 \Longleftrightarrow \mathbf{y}^1 - \mathbf{y}^2 \in \operatorname{int} D.$$

1.3 Scalarization

In order to obtain Pareto solutions, *scalarization techniques* are commonly applied. Since the objective function in multiobjective optimization is vector-valued, it is transformed into a scalar valued one. Then such a scalarization function is required to posses the following property:

Theorem 1.1. *Let F be a scalarization function which transforms a vector valued objective function $\mathbf{y} = \boldsymbol{f}(\boldsymbol{x})$ into a scalar valued one. If F preserves the Pareto order in $\mathbf{y}$, namely if F satisfies for any $\mathbf{y}^1, \mathbf{y}^2 \in \boldsymbol{f}(X)$*

$$\mathbf{y}^1 \leq \mathbf{y}^2 \Rightarrow F(\mathbf{y}^1) < F(\mathbf{y}^2), \tag{1.1}$$

then a solution $\boldsymbol{x}^0$ which minimizes F over X is a Pareto solution.

Proof. If $\boldsymbol{x}^0$ is not a Pareto solution, then there exist some $\hat{\boldsymbol{x}} \in X$ such that

$$\boldsymbol{f}(\hat{\boldsymbol{x}}) \leq \boldsymbol{f}(\boldsymbol{x}^0).$$

Since F preserves the Pareto order, we have

$$F(\boldsymbol{f}(\hat{\boldsymbol{x}})) < F(\boldsymbol{f}(\boldsymbol{x}^0)).$$

This contradicts that $\boldsymbol{x}^0$ minimizes $F(\boldsymbol{x})$ over X. $\square$

Theorem 1.2. *If F preserves the order $<$ in $\mathbf{y}$, namely if F satisfies for any $\mathbf{y}^1, \mathbf{y}^2 \in \boldsymbol{f}(X)$*
$$\mathbf{y}^1 < \mathbf{y}^2 \Rightarrow F(\mathbf{y}^1) < F(\mathbf{y}^2),$$
then a solution $\boldsymbol{x}^0$ which minimizes F over X is a weak Pareto solution.

Proof. An exercise for readers. $\square$

Theorem 1.3. *In Theorem 1.2, if F is continuous with respect to $\mathbf{y}(= \boldsymbol{f}(\boldsymbol{x}))$, and if $\boldsymbol{x}^0$ minimizing F over X is unique, then $\boldsymbol{x}^0$ is a Pareto solution.*

Proof. If $\boldsymbol{x}^0$ is not a Pareto solution, then there exists some $\hat{\boldsymbol{x}} \in X$, $\hat{\boldsymbol{x}} \neq \boldsymbol{x}^0$, such that

$$\boldsymbol{f}(\hat{\boldsymbol{x}}) \leq \boldsymbol{f}(\boldsymbol{x}^0). \tag{1.2}$$

Since $\boldsymbol{x}^0$ is already known to be a weak Pareto solution due to Theorem 1.2, there exists some i such that

$$f_i(\hat{\boldsymbol{x}}) = f_i(\boldsymbol{x}^0).$$

Let I be a set of such indices. Then (1.2) is reduced to

$$f_i(\hat{\boldsymbol{x}}) = f_i(\boldsymbol{x}^0), \quad i \in I,$$
$$f_j(\hat{\boldsymbol{x}}) < f_j(\boldsymbol{x}^0), \quad j \in \{1, \dots, r\} \backslash I.$$

Therefore, for any $\epsilon > 0$ we have

$$f_i(\hat{\boldsymbol{x}}) < f_i(\boldsymbol{x}^0) + \epsilon, \quad i \in I,$$
$$f_j(\hat{\boldsymbol{x}}) < f_j(\boldsymbol{x}^0), \qquad j \in \{1, \dots, r\} \backslash I.$$

Denote $\boldsymbol{f}^0 = \boldsymbol{f}(\boldsymbol{x}^0)$. Letting $\boldsymbol{f}^0 + \epsilon$ be a vector whose ith element is $f_i(\boldsymbol{x}^0) + \epsilon$ ($i \in I$) and jth element is $f_j(\boldsymbol{x}^0)$, $j \in \{1, \dots, r\} \backslash I$, we have

$$F(\boldsymbol{f}(\hat{\boldsymbol{x}})) < F(\boldsymbol{f}^0 + \epsilon)$$

because F preserves the order $<$ with respect to $\boldsymbol{y} (= \boldsymbol{f}(\boldsymbol{x}))$. Now, it follows from the continuity of F that

$$F(\boldsymbol{f}(\hat{\boldsymbol{x}})) \leqq F(\boldsymbol{f}(\boldsymbol{x}^0))$$

by letting $\epsilon \to 0$. Since $\boldsymbol{x}^0$ minimizes F over X, we have $\hat{\boldsymbol{x}}$ also minimizes F over X. This contradicts the uniqueness of $\boldsymbol{x}^0$ minimizing F over X. Finally, it follows that a unique solution $\boldsymbol{x}^0$ minimizing F over X is a Pareto solution.

$\square$

Remark 1.4. In terms of domination sets, the order preservation can be interpreted as follows: A scalarization function F induces a weak order $\leqq_F$. The domination set for $\leqq_F$ is given by

$$D_F^+(\boldsymbol{y}) = \{\, \boldsymbol{d} \in \mathbb{R}^r \mid \boldsymbol{y} + \boldsymbol{d} \leqq_F \boldsymbol{y} \,\}.$$

The fact that the scalarization function F preserves the Pareto order with the property of (1.1) is equivalent to that for any $\boldsymbol{y}$ in Fig. 1.4, we have

$$D^+(\boldsymbol{y}) \subset D_F^+(\boldsymbol{y}).$$

Here, $D^+(\boldsymbol{y})$ and $D_F^+(\boldsymbol{y})$ have no common boundary except for $\boldsymbol{y}$. At this event, if $D^+(\boldsymbol{y})$ and $D_F^+(\boldsymbol{y})$ have some common boundary except for $\boldsymbol{y}$, then we have

$$\boldsymbol{y}^1 \leq \boldsymbol{y}^2 \Rightarrow F(\boldsymbol{y}^1) \leqq F(\boldsymbol{y}^2).$$

1.3 Scalarization

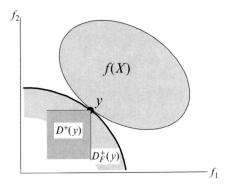

Fig. 1.4 Order preserving scalarization function

We shall discuss several typical scalarization functions in the following.

1.3.1 Linearly Weighted Sum

$$F_1(\boldsymbol{f}(\boldsymbol{x})) = w_1 f_1(\boldsymbol{x}) + \cdots + w_r f_r(\boldsymbol{x}).$$

It is clear that above F_1 preserves the Pareto order with respect to $\boldsymbol{y}\,(=\boldsymbol{f}(\boldsymbol{x}))$ for any $\boldsymbol{w}>0$. Therefore, solutions minimizing F_1 over X for any $\boldsymbol{w} > 0$ are Pareto optimal. However, F_1 does not preserve the Pareto order for any $\boldsymbol{w} \geq 0$ but preserve only the order $<$, and hence solutions minimizing F_1 over X for $\boldsymbol{w} \geq 0$ are just weakly Pareto optimal.

It is important whether the reverse of Theorem 1.1 holds in actual cases. As is readily seen in Fig. 1.5, when $\boldsymbol{f}(X)+\mathbb{R}_+^r$ is nonconvex, we cannot obtain

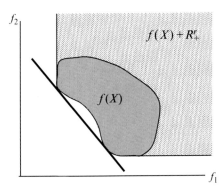

Fig. 1.5 Pareto solutions in nonconvex parts cannot be obtained by any linearly weighted sum scalarization function

Pareto solutions in a sunken (nonconvex) part by minimizing F_1 whatever the weight $w \geq 0$ may be. Hereafter, we say the Pareto frontier to be nonconvex when $f(X) + \mathbb{R}_+^r$ is nonconvex. On the other hand, the Pareto frontier is said to be convex when $f(X) + \mathbb{R}_+^r$ is convex.

1.3.2 Tchebyshev Scalarization Function

As stated above, Pareto solutions in nonconvex parts cannot be obtained by any linearly weighted sum scalarization function. It is desirable to get a Pareto solution whatever it may be by adjusting parameter w_i appropriately. It is obvious that a scalarization function with this property is the one given in Fig. 1.6. The following *weighted Tchebyshev scalarization function* holds this property:
$$F_\infty(f(x)) = \max_{1 \leq i \leq r} w_i f_i(x).$$

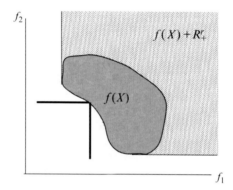

Fig. 1.6 Tchebyshev scalarization function

It can be readily seen that the above weight Tchebyshev scalarization function preserves the order $<$. Therefore, a minimal solution to F_∞ is weakly Pareto optimal (Fig. 1.6).

1.3.3 Augmented Tchebyshev Scalarization Function

It is possible to obtain any Pareto solution even in nonconvex parts by minimizing the weighted Tchebyshev scalarization function by adjusting the weight appropriately. On the other hand, however, the resulting solution is simply guaranteed to be weakly Pareto optimal. In order to guarantee the

1.3 Scalarization

Pareto optimality of the obtained solution, the following *augmented Tchebyshev scalarization function* has been developed:

$$\tilde{F}_\infty(\boldsymbol{f}(\boldsymbol{x})) = \max_{1 \leq i \leq r} w_i f_i(\boldsymbol{x}) + \alpha \sum_{i=1}^{r} w_i f_i(\boldsymbol{x}).$$

The contour of augmented Tchebyshev scalarization function is intermediate between the weighted Tchebyshev scalarization and linearly weighted sum (Fig. 1.7). The parameter α decides the angle between two half lines. If $\alpha > 0$, then $\tilde{F}_\infty$ preserves the Pareto order. Therefore, although we obtain Pareto solutions by minimizing $\tilde{F}_\infty$ for any α, we usually take a sufficiently small α (say, 10^{-6}) in order to get any Pareto solution even in nonconvex parts.

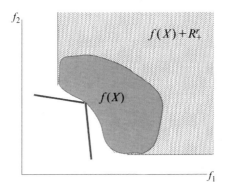

Fig. 1.7 Augmented Tchebyshev scalarization function

Remark 1.5. Since the augmented Tchebyshev scalarization function is not smooth, we cannot apply gradient based optimization methods for minimizing it. In order to overcome this difficulty, we usually solve the following equivalent problem:

$$\begin{aligned}
\underset{\boldsymbol{x},\, z}{\text{minimize}} \quad & z + \sum_{i=1}^{r} w_i f_i(\boldsymbol{x}) \quad &\text{(AT)} \\
\text{subject to} \quad & w_i f_i(\boldsymbol{x}) \leq z, \ i = 1, \ldots, r, \\
& \boldsymbol{x} \in X.
\end{aligned}$$

If the set X is open or the whole space $\mathbb{R}^n$ itself, and if $f_i(\boldsymbol{x})$, $i = 1, \ldots, r$, are appropriately smooth, then we have the following necessary condition for the solution to (AT). For Lagrange function

$$L(\boldsymbol{x}, z, \boldsymbol{\lambda}) = z + \alpha \sum_{i=1}^{r} w_i f_i(\boldsymbol{x}) + \sum_{i=1}^{r} \lambda_i (w_i f_i(\boldsymbol{x}) - z),$$

we have at the solution $\boldsymbol{x}^*,\ z^*,\ \lambda_i^*$

$$\nabla_x L = \sum_{i=1}^{r}(\alpha + \lambda_i^*)w_i \nabla f_i(\boldsymbol{x}^*) \ = \ \boldsymbol{0}, \tag{1.3}$$

$$\frac{\partial L}{\partial z} = 1 - \sum_{i=1}^{r}\lambda_i^* \ = \ 0 \ \Rightarrow \ \sum_{i=1}^{r}\lambda_i^* = 1,$$

$$\lambda_i^* \ \geqq \ 0,$$

$$w_i f_i(\boldsymbol{x}^*) \ \leqq \ z^*,$$

$$\lambda_i^*(w_i f_i(\boldsymbol{x}^*) - z^*) \ = \ 0, \ \ i = 1,\dots,r.$$

If the Pareto frontier in the objective space is smooth, we can interpret that $((\alpha + \lambda_1^*)w_1,\dots,(\alpha+\lambda_r^*)w_r)$ provides the information of normal vector of the tangent hyperplane for the Pareto frontier at $\boldsymbol{f}(\boldsymbol{x}^*)$ due to (1.3). This fact is utilized for *trade-off analysis* later in this book.

1.3.4 Constraint Transformation Method

The *constraint transformation method*[2] makes scalarization by treating one of objectives as objective function and others as constraints. Namely, without loss of generality we treat f_r as an objective function and others $f_1,\dots,f_{r-1}$ as inequality constraints as follows:

$$\begin{aligned} &\underset{\boldsymbol{x}}{\text{minimize}} && f_r(\boldsymbol{x}) && (\mathrm{S}_\varepsilon)\\ &\text{subject to} && f_i(\boldsymbol{x}) \leq \varepsilon_i, \ \ i = 1,\dots,r-1,\\ &&& \boldsymbol{x} \in X. \end{aligned}$$

Those $\varepsilon_1,\dots,\varepsilon_{r-1}$ can be regarded as *aspiration levels* for $f_1,\dots,f_{r-1}$, respectively (Fig. 1.8).

Theorem 1.4. *If $\boldsymbol{x}^0 \in X$ is an optimal solution to (S_ε) for some $\boldsymbol{\varepsilon} \in \mathbb{R}^{r-1}$, then $\boldsymbol{x}^0$ is a weak Pareto solution to (MOP). In addition, if $\boldsymbol{x}^0$ is a unique optimal solution to (S_ε), then $\boldsymbol{x}^0$ is a Pareto solution to (MOP).*

Proof. Suppose that an optimal solution $\boldsymbol{x}^0 \in X$ for $\boldsymbol{\varepsilon} \in \mathbb{R}^{r-1}$ is not a weak Pareto solution to (MOP), then there exists $\boldsymbol{x}' \in X$ such that $\boldsymbol{f}(\boldsymbol{x}') < \boldsymbol{f}(\boldsymbol{x}^0)$. This implies

$$f_r(\boldsymbol{x}') < f_r(\boldsymbol{x}^0),$$

$$f_i(\boldsymbol{x}') < f_i(\boldsymbol{x}^0) \leqq \varepsilon_i, \ \ i = 1,\dots,r-1,$$

---

[2] Haimes et al. [52] call this ε-constraint method.

1.4 Scalarization and Trade-off Analysis

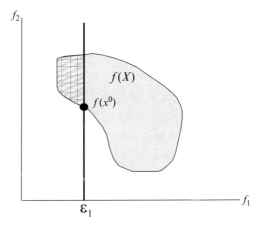

Fig. 1.8 Constraint transformation method

which contradicts that $x^0 \in X$ is an optimal solution to (S_ε). In addition, if $x^0 \in X$ is a unique optimal solution to (S_ε), then $x^0 \in X$ is Pareto optimal due to Theorem 1.3. □

1.4 Scalarization and Trade-off Analysis

Suppose that there exists a smooth function $y_r = y_r(y_1, \ldots, y_{r-1})$ in some neighborhood of a Pareto value $\boldsymbol{y}^0$ such that $\boldsymbol{y} = (y_1, \ldots, y_r)^T$ is on the Pareto frontier if and only if $y_r = y_r(y_1, \ldots, y_{r-1})$. Then

$$t_i = -\frac{\partial y_r(y_1^0, \ldots, y_{r-1}^0)}{\partial y_i}, \quad i = 1, \ldots, r-1,$$

is called the *trade-off ratio* of f_i with respect to f_r at $\boldsymbol{y}^0$. Intuitively speaking, the trade-off ratio t_i implies how much f_r can be improved in compensation of the sacrifice of f_i by a marginal unit.

In this section, we consider how we can know the value of trade-off ratio when obtaining a Pareto solution based on scalarization techniques. We shall discuss this along the constraint transformation technique since a similar derivation is possible for other scalarization methods. When the constraint set X is given by $X = \{\, \boldsymbol{x} \in \mathbb{R}^n \mid g_j(\boldsymbol{x}) \leqq 0, \ j = 1, \ldots, m \,\}$, the scalarized problem is given by

$$\begin{aligned}
\underset{\boldsymbol{x}}{\text{minimize}} \quad & f_r(\boldsymbol{x}) & (S_\varepsilon)\\
\text{subject to} \quad & g_j(\boldsymbol{x}) \leqq 0, \ j = 1, \ldots, m,\\
& f_i(\boldsymbol{x}) \leqq \varepsilon_i, \ i = 1, \ldots, r-1.
\end{aligned}$$

If a point $\boldsymbol{y} = (y_1, \ldots, y_r)^T$ is on the Pareto frontier, then under some appropriate assumption of constraint qualification there exist $\boldsymbol{x}$, $\boldsymbol{\mu}$, $\boldsymbol{\lambda}$ and $\boldsymbol{\varepsilon}$ such that

$$\nabla f_r(\boldsymbol{x}) + \sum_{i=1}^{r-1} \lambda_i \nabla f_i(\boldsymbol{x}) + \sum_{j=1}^{m} \mu_j \nabla g_j(\boldsymbol{x}) = \mathbf{0}, \tag{1.4}$$

$$\mu_j g_j(\boldsymbol{x}) = 0, \quad j = 1, \ldots, m, \tag{1.5}$$

$$\lambda_i (f_i(\boldsymbol{x}) - \varepsilon_i) = 0, \quad i = 1, \ldots, r-1, \tag{1.6}$$

$$f_i(\boldsymbol{x}) - y_i = 0, \quad i = 1, \ldots, r, \tag{1.7}$$

$$\mu_j \geqq 0, \quad g_j(\boldsymbol{x}) \leqq 0, \quad j = 1, \ldots, m,$$

$$\lambda_i \geqq 0, \quad f_i(\boldsymbol{x}) \leqq \varepsilon_i, \quad j = 1, \ldots, r-1.$$

Suppose that $\boldsymbol{x}^0$ is a solution to the problem (S_{ε^0}), and $\boldsymbol{\mu}^0$ is a Lagrange multiplier associated for the constraint $\boldsymbol{g}(\boldsymbol{x}) \leqq \mathbf{0}$ and $\boldsymbol{\lambda}^0$ the one for $\boldsymbol{f}(\boldsymbol{x}) \leqq \varepsilon^0$. Since there are $n + m + 2r - 1$ equations for (1.4)–(1.7), we try to represent the variables $(\boldsymbol{x}, \boldsymbol{\mu}, \boldsymbol{\lambda}, \boldsymbol{\varepsilon}, y_r)$ by the remainder $(y_1, \ldots, y_{r-1})$. To this end, letting the Lagrange function be denoted by

$$L(\boldsymbol{x}, \boldsymbol{\mu}, \boldsymbol{\lambda}) = f_r(\boldsymbol{x}) + \sum_{i=1}^{r-1} \lambda_i f_i(\boldsymbol{x}) + \sum_{j=1}^{m} \mu_j g_j(\boldsymbol{x}),$$

we can apply the implicit function theorem if

$$M = \begin{bmatrix}
\nabla_{xx}^2 L & \nabla^T g_1 & \cdots & \nabla^T g_m & \nabla^T f_1 & \cdots & \nabla^T f_{r-1} & \mathbf{0} & & \mathbf{0} \\
\mu_1 \nabla g_1 & g_1 & & & & & & & & \\
\vdots & & \ddots & & \mathbf{0} & & \mathbf{0} & & \mathbf{0} \\
\mu_m \nabla g_m & & & g_m & & & & & & \\
\lambda_1 \nabla f_1 & & & & f_1 - \varepsilon_1 & & & -\lambda_1 & & \\
\vdots & & \mathbf{0} & & & \ddots & & & \ddots & \\
\lambda_{r-1} \nabla f_{r-1} & & & & & & f_{r-1} - \varepsilon_{r-1} & & -\lambda_{r-1} & \\
\nabla f_1 & & & & & & & & & \\
\vdots & & \mathbf{0} & & & \mathbf{0} & & & \mathbf{0} & \mathbf{0} \\
\nabla f_r & & & & & & & & & -1
\end{bmatrix}$$

is nonsingular at $(\boldsymbol{x}^0, \boldsymbol{\mu}^0, \boldsymbol{\lambda}^0, \varepsilon^0, \boldsymbol{y}^0)$. In order to check whether this condition holds, we can apply the well known sensitivity analysis in mathematical programming (see, e.g., Fiacco [43]) as follows:

1.4 Scalarization and Trade-off Analysis

Let $J(\boldsymbol{x}^0)$ denote the set of indices of g_j such that $g_j(\boldsymbol{x}^0) = 0$:

I. Linear independence

$$\{ \nabla f_i(\boldsymbol{x}^0), \ i = 1, \ldots, r-1; \ \nabla g_j(\boldsymbol{x}^0), \ j \in J(\boldsymbol{x}^0) \}$$

are linearly independent.

II. Strictly complementary slackness

(a) $\lambda_i^0 > 0$ (hence $f_i(\boldsymbol{x}^0) - \varepsilon_i^0 = 0, \ i = 1, \ldots, r-1$)

(b) $\mu_j^0 > 0, \ (j \in J(\boldsymbol{x}^0))$

III. Second order optimality condition

For any $\boldsymbol{u} \neq \boldsymbol{0}$ satisfying

$$\nabla f_i(\boldsymbol{x}^0)\boldsymbol{u} = 0, \ i = 1, \ldots, r-1; \ \nabla g_j(\boldsymbol{x}^0)\boldsymbol{u} = 0, \ j \in J(\boldsymbol{x}^0),$$

we have

$$\boldsymbol{u}^T \nabla_{xx}^2 L(\boldsymbol{x}^0, \boldsymbol{\mu}^0, \boldsymbol{\lambda}^0, \boldsymbol{\varepsilon}^0)\boldsymbol{u} > 0.$$

If the conditions I–III hold, $(\boldsymbol{x}, \ \boldsymbol{\mu}, \ \boldsymbol{\lambda}, \ \boldsymbol{\varepsilon}, \ y_r)$ can be expressed by a first order continuously differentiable function in terms of $(y_1, \ldots, y_{r-1})$. Furthermore, the derivative of (1.4)–(1.7) with respect to $(y_1, \ldots, y_{r-1})$ is given by

$$\begin{bmatrix} \boldsymbol{0} \\ \boldsymbol{0} \\ -\boldsymbol{I}_{r-1} \\ \boldsymbol{0} \end{bmatrix}.$$

Letting the last row vector of M^{-1} be

$$\left[\boldsymbol{u}^T, \ \boldsymbol{v}^T, \ \boldsymbol{s}^T, \ -\boldsymbol{t}^T \right], \ \boldsymbol{u} \in \mathbb{R}^n, \ \boldsymbol{v} \in \mathbb{R}^m, \ \boldsymbol{s} \in \mathbb{R}^{r-1}, \ \boldsymbol{t} \in \mathbb{R}^r,$$

we have

$$\frac{\partial y_r(y_1, \ldots, y_{r-1})}{\partial y_i} = -t_i, \ i = 1, \ldots, r-1.$$

Note that at $(\boldsymbol{x}^0, \ \boldsymbol{\mu}^0, \ \boldsymbol{\lambda}^0, \ \varepsilon^0, \ \boldsymbol{y}^0)$, we have

$$\boldsymbol{u} = \boldsymbol{0},$$

$$v_j = \begin{cases} -1, & j \in J(\boldsymbol{x}^0) \\ 0, & j \notin J(\boldsymbol{x}^0) \end{cases},$$

$$\boldsymbol{s} = \boldsymbol{0},$$

$$t_i = \lambda_i^0, \quad i = 1, \ldots, r-1,$$

$$t_r = 1.$$

Therefore, it follows that the trade-off ratio of f_i with respect to f_r at $\boldsymbol{y}^0$ can be obtained as the Lagrange multiplier λ_i^0 associated with the constraint $f_i(\boldsymbol{x}) \leqq \varepsilon_i^0$ for the problem (S_{ε^0}).

Applying the same technique to the linearly weighted sum scalarization function, we have

$$t_i = \frac{w_i}{w_r}, \quad i = 1, \ldots, r-1.$$

Similarly, for Tchebyshev scalarization function

$$t_i = \frac{\lambda_i w_i}{\lambda_r w_r}, \quad i = 1, \ldots, r-1,$$

where λ_i is the Lagrange multiplier associated with the constraint $w_i f_i(\boldsymbol{x}) \leqq z$.

Remark 1.6. If the function $y_r = y_r(y_1, \ldots, y_{r-1})$ representing the Pareto frontier in the objective space is once continuously differentiable, we have the following form of total differential:

$$dy_r = -t_1 dy_1 - \cdots - t_{r-1} dy_{r-1}.$$

Therefore, it follows for the Tchebyshev scalarization function that

$$\lambda_1 w_1 dy_1 + \cdots + \lambda_{r-1} w_{r-1} dy_{r-1} + \lambda_r w_r dy_r = 0. \tag{1.8}$$

This relation is often used in actual decision making as the trade-off relation for all objective functions, when we consider changes among several objectives simultaneously.

Remark 1.7. It is often difficult to check whether the conditions I–III hold in many actual cases. For convex cases, however, it is relatively easier to make the trade-off analysis. We shall explain this below along the Tchebyshev scalarization function. Suppose that the objective functions $f_i, \ i = 1, \ldots, r$, and the constraint functions $g_j, \ j = 1, \ldots, m$, are convex. Let $\boldsymbol{x}^0$ be a solution to the scalarized optimization problem for the Tchebyshev scalarization function. Then for the Lagrange function L given by

$$L(z, \boldsymbol{x}, \boldsymbol{\lambda}) = z + \sum_{i=1}^{r} \lambda_i (w_i f_i(\boldsymbol{x}) - z),$$

we have

$$L(z^0, \boldsymbol{x}^0, \boldsymbol{\lambda}) \leqq L(z, \boldsymbol{x}, \boldsymbol{\lambda}) \quad \text{for } \forall z, \ \forall \boldsymbol{x} \in X, \tag{1.9}$$

$$\lambda_i (w_i f_i(\boldsymbol{x}) - z) = 0, \quad \lambda_i \geqq 0, \quad i = 1, \ldots, r$$

(see, e.g., Lasdon [79]).

1.4 Scalarization and Trade-off Analysis

Since (1.9) holds for any z, we have by differentiating it with respect to z

$$\sum_{i=1}^{r} \lambda_i = 1.$$

Rewriting (1.9) with the above relation yields

$$\sum_{i=1}^{r} \lambda_i w_i f_i(\boldsymbol{x}^0) \leqq \sum_{i=1}^{r} \lambda_i w_i f_i(\boldsymbol{x}) \ \text{ for } \forall x \in X,$$

which implies that $(\lambda_1 w_1, \ldots, \lambda_r w_r)$ is the normal vector of the supporting hyperplane of the set $Y(=\boldsymbol{f}(X))$ at $\boldsymbol{y}^0(=\boldsymbol{f}(\boldsymbol{x}^0))$. In other words, $(\lambda_1 w_1/\lambda_r w_r, \ldots, \lambda_{r-1} w_{r-1}/\lambda_r w_r)$ is a subgradient of the function $y_r = y_r(y_1, \ldots, y_{r-1})$ representing the Pareto frontier. Therefore, it can be regarded that (1.8) provides the trade-off relation also in convex cases, although its precision may not be so good. If the above Lagrange multiplier is unique, then (1.8) gives the perfect linear approximation of Pareto frontier, namely the exact information of trade-off (see the following lemma).

Lemma 1.1. *A convex function $F(\boldsymbol{y})$ is differentiable at $\hat{\boldsymbol{y}}$ if and only if $F(\boldsymbol{y})$ possesses a unique subgradient at $\hat{\boldsymbol{y}}$ (Rockafellar [125]).*

Remark 1.8. As a special case, consider the case in which the set X is convex polyhedral and all f_i, $i = 1, \ldots, r$, are linear.

Lemma 1.2. *Let $\boldsymbol{f} = (f_1, \ldots, f_r)^T$. If all f_i, $i = 1, \ldots, r$ are linear and if a set X is convex polyhedral, then the set $\boldsymbol{f}(X)$ is also convex polyhedral (Rockafellar [125]).*

It is known that the Pareto frontier P_L for linear case is convex (namely, $P_L + \mathbb{R}_+^r$ is convex) and moreover piecewise linear. In this case, since the auxiliary scalarized optimization problem (S_ε) is linear, the Lagrange multiplier is unique for nondegenerated solution. Therefore, (1.8) provides a complete information of trade-off relation among the objectives. If the solution to the auxiliary scalarized problem is degenerated, then the Lagrange multiplier is not unique. Even in this case, we can make the trade-off analysis easily by using parametric optimization technique (see for more details Nakayama [98]).

Chapter 2
Interactive Programming Methods for Multiobjective Optimization

In general, there may be many Pareto solutions in multiobjective optimization problems. The final decision is made among them taking the total balance over all objectives into account. This is a problem of value judgment of decision maker (DM). The totally balancing over criteria is usually called *trade-off*. Interactive multiobjective programming searches a solution in an interactive way with DM while eliciting information on his/her value judgment. Then it is important how easily DM can make trade-off analysis to get a final solution. To this aim, several kinds of interactive techniques for multiple criteria decision making have been developed so far. For details, see the literatures [47, 90, 130, 147, 157, 159]. Above all, the *aspiration level approach* (reference point methods in some literatures) is now widely recognized to be effective in many practical fields, because:

1. It does not require any consistency of DM's judgment.
2. Aspiration levels reflect the wish of DM very well.
3. Aspiration levels play the role of probe better than the weight for objective functions.

In this chapter, first we will discuss the difficulty in weighting method which is commonly used in the traditional goal programming, and next explain how the aspiration level approach overcomes this difficulty.

2.1 Goal Programming

Goal programming (GP) was proposed by Charnes and Cooper in 1961 to get rid of cases in which no feasible solution exists in usual linear programming (LP) [17]. In order to explain GP, we should begin with the story of birth of LP.

H. Nakayama et al., *Sequential Approximate Multiobjective Optimization*
Using Computational Intelligence, Vector Optimization,
DOI 10.1007/978-3-540-88910-6_2, © 2009 Springer-Verlag Berlin Heidelberg

Dantzig [28] retrospected the birth of LP as follows: "Initially, there was no objective function: explicit *goals* did not exist because practical planners simply had no way to implement such a concept. ... By mid-1947 I decided that the objective had to be made explicit. ... The use of linear form as the objective function to be extremized was the novel feature. ..."

LP has been applied to various kinds of practical problems since its proposal. Practitioners, in particular engineers, however, often encounter cases in which no feasible solution exists. The motivation that Dantzig introduced the objective function to be maximized seems to owe to an idea of utilitarianism that human beings try to maximize their utilities on the background. On the contrary, Simon insisted that the rationality of human behavior is in "satisficing" rather than in optimization [142].

In order to get rid of cases in which no feasible solution exists in LP, Charnes and Cooper introduced the idea of goal attainment along the idea of satisficing: they insisted that any constraints can be regarded as "goal."

Goals in mathematical programming model are classified into the following three categories:

1. $g_i(\boldsymbol{x}) \geqq \overline{g}_i, \quad i \in I_{GT},$
2. $g_i(\boldsymbol{x}) \leqq \overline{g}_i, \quad i \in I_{LT},$
3. $g_i(\boldsymbol{x}) = \overline{g}_i, \quad i \in I_{EQ},$

where $\overline{g}_i$ is called the *goal level* of g_i. The goal itself may be feasible or infeasible. In cases where the goal is infeasible, the aim of goal programming is to obtain a solution as close to the given goal level as possible.

Let

$$g(\boldsymbol{x}) + y^+ - y^- = \overline{g}, \quad y^+ \geqq 0, \ y^- \geqq 0.$$

Then the amount y^+ and y^- denotes, respectively, the *overattainment* and the *underattainment*, if $y^+y^- = 0$ and less of the level of the criterion g is preferred to more.

Now, GP is formulated in general as follows:

$$\begin{aligned}
\underset{\boldsymbol{x},\,\boldsymbol{y}^+,\,\boldsymbol{y}^-}{\text{minimize}} \quad & \sum_{i \in I_{GT}} w_i y_i^+ + \sum_{i \in I_{LT}} w_i y_i^- + \sum_{i \in I_{EQ}} w_i(y_i^+ + y_i^-) \quad \text{(GP)} \\
\text{subject to} \quad & g_i(\boldsymbol{x}) + y_i^+ - y_i^- = \overline{g}_i, \quad i \in I_{GT} \cup I_{LT} \cup I_{EQ}, \\
& y_i^+, \ y_i^- \geqq 0, \quad i \in I_{GT} \cup I_{LT} \cup I_{EQ}, \\
& \boldsymbol{x} \in X \subset \mathbb{R}^n.
\end{aligned}$$

Remark 2.1. The condition $y_i^+ y_i^- = 0$ is crucial for y_i^+ and y_i^- to have the meaning of the overattainment and the underattainment, respectively, if less of the level of the criterion g_i is preferred to more. Fortunately, the optimal solution to (GP) satisfies this condition automatically. The following lemma proves this in a more general form.

2.1 Goal Programming

Lemma 2.1. *Let $\boldsymbol{y}^+$ and $\boldsymbol{y}^-$ be vectors of $\mathbb{R}^r$. Then consider the following problem:*

$$
\begin{array}{ll}
\underset{\boldsymbol{x},\,\boldsymbol{y}^+,\,\boldsymbol{y}^-}{minimize} & G(\boldsymbol{y}^+, \boldsymbol{y}^-) \\[2mm]
subject\ to & g_i(\boldsymbol{x}) + y_i^+ - y_i^- = \overline{g}_i, \quad i = 1,\ldots,r, \\[2mm]
& y_i^+,\ y_i^- \geqq 0, \quad i = 1,\ldots,r, \\[2mm]
& \boldsymbol{x} \in X \subset \mathbb{R}^n.
\end{array}
$$

Suppose that the function G is monotonically increasing with respect to elements of $\boldsymbol{y}^+$ and $\boldsymbol{y}^-$ and strictly monotonically increasing with respect to at least either y_i^+ or y_i^- for each i, $i = 1,\ldots,r$. Then, the solution $\hat{\boldsymbol{y}}^+$ and $\hat{\boldsymbol{y}}^-$ to the above problem satisfy

$$
\hat{y}_i^+ \hat{y}_i^- = 0, \qquad i = 1,\ldots,r.
$$

Proof. See, e.g., [130]. $\square$

Consider a case in which less values of the criteria f_i, $i = 1,\ldots,r$ are more preferable, but it is desirable for f_i to be at least less than $\overline{f}_i$, $i = 1,\ldots,r$. For this situation, GP is formulated as follows:

$$
\begin{array}{ll}
\underset{\boldsymbol{x},\,\boldsymbol{y}^+,\,\boldsymbol{y}^-}{minimize} & \displaystyle\sum_i^r w_i y_i^- \qquad\qquad\qquad\qquad (\text{GP}') \\[4mm]
subject\ to & f_i(\boldsymbol{x}) + y_i^+ - y_i^- = \overline{f}_i, \quad i = 1,\ldots,r, \\[2mm]
& y_i^+,\ y_i^- \geqq 0, \quad i = 1,\ldots,r, \\[2mm]
& \boldsymbol{x} \in X \subset \mathbb{R}^n.
\end{array}
$$

According to Lemma 2.1, we have $y_i^+ y_i^- = 0$ for $i = 1,\ldots,r$ at the optimal solution. In the above problem (GP$'$), therefore, y_i^- has the meaning of underattainment of the criterion f_i. Since $w_i \geqq 0$, $y_i^+ \geqq 0$, $y_i^- \geqq 0$, for $i = 1,\ldots,r$, the optimal value of objective function of (GP$'$) is nonnegative. If some y_i^- is positive at the solution, the goal level $\overline{f}_i$ is not attained, while if $y_i^- = 0$ at the solution, the goal level $\overline{f}_i$ is attained. In cases with $y_i^- = 0$, it should be noted that we have a solution $\hat{\boldsymbol{x}}$ with $f_i(\hat{\boldsymbol{x}}) \leqq \overline{f}_i$ but cannot improve f_i any more than the level $f_i(\hat{\boldsymbol{x}})$, even though there are feasible solutions which yield less values than $f_i(\hat{\boldsymbol{x}})$. In other words, goal programming does not necessarily assure the Pareto optimality of solution. This seems reasonable because the basic idea of goal programming is based on satisficing. However, this fact is not satisfactory in view of criteria for which less values are more preferable. It is not so difficult to find any solution which yields as less value of f_i as possible keeping less than $\overline{f}_i$, if any, for the problems which can be formulated as mathematical programming including goal programming. Therefore, this fact is one of drawbacks of goal programming.

Remark 2.2. If less level of f is more preferable, and if we aim to find a solution which yields as less level of f as possible even though it clears the goal level $\overline{f}$, then we should formulate as follows:

$$\begin{array}{ll} \underset{\boldsymbol{x},\,z}{\text{minimize}} & \sum_{i=1}^{r} w_i z_i \qquad\qquad\qquad (\text{GP}'') \\ \text{subject to} & f_i(\boldsymbol{x}) - \overline{f}_i \leqq z_i, \quad i = 1,\ldots,r, \\ & \boldsymbol{x} \in X \subset \mathbb{R}^n. \end{array}$$

Note that the new variable z is not necessarily nonnegative. The stated problem is equivalent to multiobjective programming problem:

$$\begin{array}{ll} \underset{\boldsymbol{x}}{\text{minimize}} & \sum_{i=1}^{r} w_i f_i(\boldsymbol{x}) \\ \text{subject to} & \boldsymbol{x} \in X \subset \mathbb{R}^n. \end{array}$$

However, by (GP''), we can treat f_i not only as an objective function but also as a constraint by setting $z_i = 0$. This technique will be used to interchange the role of objective function and constraint in subsequent sections.

From a viewpoint of decision making, it is important to obtain a solution which the decision maker accepts easily. This means it is important whether the obtained solution is suited to the value judgment of the decision maker. In goal programming, the adjustment of weights w_i is required to obtain a solution suited to the value judgment of the decision maker. However, this task is very difficult in many cases as will be seen in the following.

2.2 Why is the Weighting Method Ineffective?

In multiobjective programming problems, the final decision is made on the basis of the value judgment of DM. Hence it is important how we elicit the value judgment of DM. In many practical cases, the vector objective function is scalarized in such a manner that the value judgment of DM can be incorporated.

The most well-known scalarization technique is the linearly weighted sum

$$\sum_{i=1}^{r} w_i f_i(x).$$

The value judgment of DM is reflected by the weight. Although this type of scalarization is widely used in many practical problems, there is a serious drawback in it. Namely, it cannot provide a solution among sunken parts of Pareto surface (frontier) due to *duality gap* for nonconvex cases. Even for

2.2 Why is the Weighting Method Ineffective?

convex cases, for example, in linear cases, even if we want to get a point in the middle of line segment between two vertices, we merely get a vertex of Pareto surface, as long as the well-known simplex method is used. This implies that depending on the structure of problem, the linearly weighted sum cannot necessarily provide a solution as DM desires.

In an extended form of goal programming, e.g., *compromise programming* [165], some kind of metric function from the goal $\overline{f}$ is used as the one representing the preference of DM. For example, the following is well known:

$$\left(\sum_{i=1}^{r} w_i |f_i(x) - \overline{f}_i|^p \right)^{1/p} . \tag{2.1}$$

The preference of DM is reflected by the weight w_i, the value of p, and the value of the goal $\overline{f}_i$. If the value of p is chosen appropriately, a Pareto solution among a sunken part of Pareto surface can be obtained by minimizing (2.1). However, it is usually difficult to predetermine appropriate values of them. Moreover, the solution minimizing (2.1) cannot be better than the goal $\overline{f}$, even though the goal is pessimistically underestimated.

In addition, one of the most serious drawbacks in the weighted sum scalarization is that people tend to misunderstand that a desirable solution can be obtained by adjusting the weight. It should be noted that there is no positive correlation between the weight w_i and the value $f(\hat{x})$ corresponding to the resulting solution $\hat{x}$ as will be seen in the following example.

Example 2.1. Let $f_1 := y_1$, $f_2 := y_2$ and $f_3 := y_3$, and let the feasible region in the objective space be given by

$$\left\{ (y_1, y_2, y_3) \mid (y_1 - 1)^2 + (y_2 - 1)^2 + (y_3 - 1)^2 \leqq 1 \right\}.$$

Suppose that the goal is $(\overline{y}_1, \overline{y}_2, \overline{y}_3) = (0, 0, 0)$. The solution minimizing the metric function (2.1) with $p = 1$ and $w_1 = w_2 = w_3 = 1$ is $(y_1, y_2, y_3) = (0.42265, 0.42265, 0.42265)$. Now suppose that DM wants to decrease the value of f_1 a lot more and that of f_2 a little more, and hence modify the weight into $w_1' = 10$, $w_2' = 2$, $w_3' = 1$. The solution associated with the new weight is $(0.02410, 0.80482, 0.90241)$. Note that the value of f_2 is worse than before despite that DM wants to improve it by increasing the weight of f_2 up to twice. Someone might think that this is due to the lack of normalization of weight. Therefore, we normalize the weight by $w_1 + w_2 + w_3 = 1$. The original weight normalized in this way is $w_1 = w_2 = w_3 = 1/3$ and the renewed weight by the same normalization is $w_1' = 10/13$, $w_2' = 2/13$, $w_3' = 1/13$. We can observe that w_2' is less than w_2. Now increase the normalized weight w_2 to be greater than $1/3$. To this end, set the unnormalized weight $w_1 = 10$, $w_2 = 7$ and $w_3 = 1$. With this new weight, we have a solution $(0.18350, 0.42845, 0.91835)$. Despite that the normalized weight $w_2'' = 7/18$ is greater than the original one $(=1/3)$, the obtained solution is still worse than the previous one.

As is readily seen in the above example, it is usually very difficult to adjust the weight in order to obtain a solution as DM wants. This difficulty seems to be caused due to the fact that there is no positive correlation between the weight and the resulting solution in cases with more than two objective functions. Therefore, it seems much better to take another probe for getting a solution which DM desires. The aspiration level of DM is promising as the probe. Interactive multiobjective programming techniques based on aspiration levels have been developed so that the drawbacks of the traditional weighting method may be overcome [13,157]. In Sect. 2.3, we shall discuss the *satisficing trade-off method* developed by one of the authors (Nakayama [100]) as an example.

2.3 Satisficing Trade-off Method

In the *aspiration level approach*, the aspiration level at the kth iteration $\overline{\boldsymbol{f}}^k$ is modified as follows:

$$\overline{\boldsymbol{f}}^{k+1} = T \circ P(\overline{\boldsymbol{f}}^k)$$

Here, the operator P selects the Pareto solution nearest in some sense to the given aspiration level $\overline{\boldsymbol{f}}^k$. The operator T is the trade-off operator which changes the kth aspiration level $\overline{\boldsymbol{f}}^k$ if DM does not compromise with the shown solution $P(\overline{\boldsymbol{f}}^k)$. Of course, since $P(\overline{\boldsymbol{f}}^k)$ is a Pareto solution, there exists no feasible solution which makes all criteria better than $P(\overline{\boldsymbol{f}}^k)$, and thus DM has to trade-off among criteria if he wants to improve some of criteria. Based on this trade-off, a new aspiration level is decided as $T \circ P(\overline{\boldsymbol{f}}^k)$. Similar process is continued until DM obtains an agreeable solution. This idea is implemented in DIDASS [51] and the satisficing trade-off method [100]. While DIDASS leaves the trade-off to the heuristics of DM, the satisficing trade-off method provides a device based on the sensitivity analysis which will be stated later.

2.3.1 On the Operation P

The operation which gives a Pareto solution $P(\overline{\boldsymbol{f}}^k)$ nearest to $\overline{\boldsymbol{f}}^k$ is performed by some *auxiliary scalar optimization*. It has been shown in [130] that the only one scalarization technique, which provides any Pareto solution regardless of the structure of problem, is of the Tchebyshev type. As was stated before, however, the Tchebyshev type scalarization function yields not only a Pareto solution but also a weak Pareto solution. Since weak Pareto solutions have a possibility that there may be another solution which improves a criteria while others being fixed, they are not necessarily "efficient" as a solution in

2.3 Satisficing Trade-off Method

decision making. In order to exclude weak Pareto solutions, we apply the *augmented Tchebyshev scalarization function*:

$$\max_{1 \le i \le r} \ w_i(f_i(\boldsymbol{x}) - \overline{f}_i) + \alpha \sum_{i=1}^{r} w_i f_i(\boldsymbol{x}), \tag{2.2}$$

where α is usually set a sufficiently small positive number, say 10^{-6}. The weight w_i is usually given as follows: Let f_i^* be an ideal value which is usually given in such a way that $f_i^* < \min\{f_i(\boldsymbol{x}) \mid \boldsymbol{x} \in X\}$, and let f_{*i} be a nadir value which is usually given by

$$f_{*i} = \max_{1 \le j \le r} \ f_i(\boldsymbol{x}_j^*),$$

where

$$\boldsymbol{x}_j^* = \arg \min_{\boldsymbol{x} \in X} f_j(\boldsymbol{x}).$$

For this circumstance, we set

$$w_i^k = \frac{1}{\overline{f}_i^k - f_i^*} \tag{2.3}$$

or

$$w_i^{k'} = \frac{1}{f_{*i} - f_i^*}. \tag{2.4}$$

The minimization of (2.2) with (2.3) or (2.4) is usually performed by solving the following equivalent optimization problem, because the original one is not smooth:

$$\begin{aligned} \underset{\boldsymbol{x}, z}{\text{minimize}} \quad & z + \alpha \sum_{i=1}^{r} w_i f_i(\boldsymbol{x}) \tag{AP} \\[2mm] \text{subject to} \quad & w_i^k \left(f_i(\boldsymbol{x}) - \overline{f}_i^k \right) \le z, \quad i = 1, \dots, r \tag{2.5} \\[2mm] & \boldsymbol{x} \in X. \end{aligned}$$

Remark 2.3. Note the weight (2.3) depends on the kth aspiration level, while the one by (2.4) is independent of aspiration levels. The difference between solutions to (AP) for these two kinds of weight can be illustrated in Fig. 2.1. In the auxiliary min–max problem (AP) with the weight by (2.3), $\overline{f}_i^k$ in the constraint (2.5) may be replaced with f_i^* without any change in the solution. For we have

$$\frac{f_i(\boldsymbol{x}) - f_i^*}{\overline{f}_i^k - f_i^*} = \frac{f_i(\boldsymbol{x}) - \overline{f}_i^k}{\overline{f}_i^k - f_i^*} + 1.$$

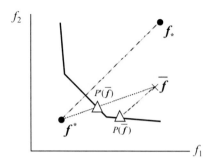

Fig. 2.1 Difference between solutions associated with two kinds of weight

The following theorem is a slight modification of Lemmas 7.3–7.5 of Sawaragi–Nakayama–Tanino [130]. It assures that we can obtain a desirable Pareto solution by adjusting the aspiration level $\overline{f}$ appropriately.

Theorem 2.1. *For arbitrary $w \geq 0$ and $\alpha > 0$, $\hat{x} \in X$ minimizing (2.2) is a properly Pareto optimal solution to (MOP). Conversely, if $\hat{x}$ is a properly Pareto optimal solution to (MOP), then there exist $\alpha > 0$ and $\overline{f}$ such that $\hat{x}$ minimizes (2.2) (or equivalently (AP)) with w decided by (2.3) or (2.4) over X.*

2.3.2 On the Operation T

In cases DM is not satisfied with the solution for $P(\overline{f}^k)$, he/she is requested to answer his/her new aspiration level $\overline{f}^{k+1}$. Let x^k denote the Pareto solution obtained by operation $P(\overline{f}^k)$, and classify the objective functions into the following three groups:

1. The class of criteria I_I^k which are to be improved more
2. The class of criteria I_R^k which may be relaxed
3. The class of criteria I_A^k which are acceptable as they are

Clearly, $\overline{f}_i^{k+1} < f_i(x^k)$ for all $i \in I_I^k$. Usually, for $i \in I_A^k$, we set $\overline{f}_i^{k+1} = f_i(x^k)$. For $i \in I_R^k$, DM has to agree to increase the value of $\overline{f}_i^{k+1}$. It should be noted that an appropriate sacrifice of f_j for $j \in I_R^k$ is needed for attaining the improvement of f_i for $i \in I_I^k$.

Remark 2.4. It is not necessarily needed to classify all of the objective functions explicitly. Indeed, in many practical cases, the objective functions are

2.3 Satisficing Trade-off Method

automatically classified after setting new aspiration levels. However, in using *automatic trade-off* or *exact trade-off* which will be stated later, we need classify criteria to be relaxed and accepted to decide their amounts of relaxation.

Example 2.2. Consider the same problem as in Example 2.1: Let $f_1 := y_1$, $f_2 := y_2$ and $f_3 := y_3$, and let the feasible region in the objective space be given by

$$\{ (y_1, y_2, y_3) \mid (y_1 - 1)^2 + (y_2 - 1)^2 + (y_3 - 1)^2 \leqq 1 \}.$$

Suppose that the ideal point is $(y_1^*, \ y_2^*, \ y_3^*) = (0, 0, 0)$, and the nadir point is $(y_{*1}, \ y_{*2}, \ y_{*3}) = (1, 1, 1)$. Therefore, using (2.4) we have $w_1 = w_2 = w_3 = 1.0$. Let the first aspiration level be $(\overline{y}_1^1, \ \overline{y}_2^1, \ \overline{y}_3^1) = (0.4, \ 0.4, \ 0.4)$. Then the solution to (AP) is $(y_1^1, \ y_2^1, \ y_3^1) = (0.423, \ 0.423, \ 0.423)$. Now suppose that DM wants to decrease the value of f_1 a lot more and that of f_2 a little more, and hence modify the aspiration level into $\overline{y}_1^2 = 0.35$ and $\overline{y}_2^2 = 0.4$. Since the present solution $(y_1^1, \ y_2^1, \ y_3^1) = (0.423, \ 0.423, \ 0.423)$ is already Pareto optimal, it is impossible to improve all of criteria. Therefore, suppose that DM agrees to relax f_3, and with its new aspiration level of $\overline{y}_3^2 = 0.5$. With this new aspiration level, the solution to (AP) is $(y_1^2, \ y_2^2, \ y_3^2) = (0.366, \ 0.416, \ 0.516)$. Although the obtained solution does not attain the aspiration level of f_1 and f_2 a little bit, it should be noted that the solution is improved more than the previous one. The reason why the improvement of f_1 and f_2 does not attain the wish of DM is that the amount of relaxation of f_3 is not much enough to compensate for the improvement of f_1 and f_2. In the satisficing trade-off method, DM can find a satisfactory solution easily by making the trade-off analysis deliberately. To this end, it is also possible to use the sensitivity analysis in mathematical programming (refer to the following automatic trade-off or exact trade-off). We have observed in the previous section that it is difficult to adjust weights for criteria so that we may get a desirable solution in the goal programming. However, the aspiration level can lead DM to his/her desirable solution easily in many practical problems.

Remark 2.5. The idea of classification and trade-off of criteria is originated from STEM [7] and followed also by NIMBUS developed by Miettinen [90].

2.3.3 Automatic Trade-off

It is of course possible for DM to answer new aspiration levels of all objective functions. In practical problems, however, we often encounter cases with a large number of objective functions, for which DM tends to get tired with answering new aspiration levels for all objective functions. Therefore, it is more practical in problems with very many objective functions for DM to answer only his/her improvement rather than both improvement and relaxation. Using the properties described in Sect. 1.4 for the augmented Tchebyshev scalarization function, we have the following relation: For some perturbation Δf_i, $i = 1, \ldots, r$ from a Pareto value,

$$0 = \sum_{i=1}^{r} (\lambda_i + \alpha) w_i \Delta f_i + o(\| \Delta f \|), \tag{2.6}$$

where λ_i is the Lagrange multiplier associated with the constraints in problem (AP). Therefore, under some appropriate condition, $((\lambda_1 + \alpha)w_1, \ldots, (\lambda_r + \alpha)w_r)$ is the normal vector of the tangent hyperplane of the Pareto surface. In particular, in multiobjective linear programming problems, the simplex multipliers corresponding to a nondegenerated solution is the feasible trade-off vector of Pareto surface [98].

Dividing the right-hand side of (2.6) into the total amount of improvement and that of relaxation,

$$0 \cong \sum_{i \in I_I} (\lambda_i + \alpha) w_i \Delta f_i + \sum_{j \in I_R} (\lambda_j + \alpha) w_j \Delta f_j.$$

Using the above relation, we can assign the amount of sacrifice for f_j ($j \in I_R$) which is automatically set in the equal proportion to $(\lambda_i + \alpha)w_i$, namely, by

$$\Delta f_j = \frac{-1}{N(\lambda_j + \alpha)w_j} \sum_{i \in I_I} (\lambda_i + \alpha) w_i \Delta f_i, \quad j \in I_R, \tag{2.7}$$

where N is the number of elements of the set I_R.

By using the above automatic trade-off method, the burden of DM can be decreased so much in cases that there are a large number of criteria. Of course, if DM does not agree with this quota Δf_j laid down automatically, he/she can modify it in a manual way.

Example 2.3. Consider the same problem as in Example 2.2. Suppose that DM has the solution $(y_1^1, y_2^1, y_3^1) = (0.423, 0.423, 0.423)$ associated with his first aspiration level $(\overline{y}_1^1, \overline{y}_2^1, \overline{y}_3^1) = (0.4, 0.4, 0.4)$. Now suppose that DM modifies the aspiration level into $\overline{y}_1^2 = 0.35$ and $\overline{y}_2^2 = 0.4$. For the amount of improvement of $|\Delta f_1| = 0.073$ and $|\Delta f_2| = 0.023$, the amount of relaxation of f_3 on the basis of automatic trade-off is $|\Delta f_3| = 0.095$. In other words, the new aspiration level of f_3 should be 0.52. If DM agrees with this trade-off, he/she will have the new Pareto solution $(y_1^2, y_2^2, y_3^2) = (0.354, 0.404, 0.524)$ to the problem (AP) corresponding to the new aspiration level $(\overline{y}_1^2, \overline{y}_2^2, \overline{y}_3^2) = (0.35, 0.4, 0.52)$. It should be noted that the obtained solution is much closer to DM's wish rather than the one in Example 2.2.

2.3 Satisficing Trade-off Method

Example 2.4. Consider the following multiple objective linear programming problem:

$$\begin{aligned}
\minimize_{x_1, x_2} \quad & f_1 = -2x_1 - x_2 + 25 \\
& f_2 = x_1 - 2x_2 + 18 \\
\text{subject to} \quad & -x_1 + 3x_2 \leqq 21, \\
& x_1 + 3x_2 \leqq 27, \\
& 4x_1 + 3x_2 \leqq 45, \\
& 3x_1 + x_2 \leqq 30, \\
& x_1, \ x_2 \geqq 0.
\end{aligned}$$

Suppose that the ideal point $\boldsymbol{f}^* = (4, \ 4)^T$ and the nadir point $\boldsymbol{f}_* = (18, \ 21)^T$ by using the payoff matrix based on minimization of each objective function separately. Letting the first aspiration level be $\overline{\boldsymbol{f}}^1 = (15, \ 9)^T$, we have the first Pareto solution $(11.632, 4.910)$ by solving the auxiliary min–max problem (AP). This Pareto point in the objective function space is the intercept of the line parallel to the line passing through $\boldsymbol{f}^*$ and $\boldsymbol{f}_*$ with the Pareto surface (curve, in this case). Now we shall consider the following three cases:

Case 1. Suppose that DM is not satisfied with the obtained Pareto solution, and he/she wants to improve the value of f_2. Let the new aspiration level of f_2 be 4.5. The relaxation amount for f_1 calculated by the automatic trade-off is 2.87. Therefore, the new aspiration level of f_1 based on the automatic trade-off is 14.5. Since the automatic trade-off is based on the linear approximation of Pareto surface at the present point, the new aspiration level obtained by the automatic trade-off is itself Pareto optimal in this case as shown in Fig. 2.2.

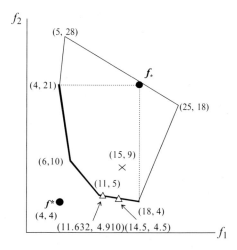

Fig. 2.2 Automatic trade-off (case 1)

Case 2. Suppose that DM wants to improve f_1 rather than f_2 at the moment when the first Pareto solution $(11.632, 4.910)$ is obtained. Let the new aspiration level of f_1 that DM desires be 9.0. Then the new aspiration level by the automatic trade-off is $\overline{\boldsymbol{f}}^2 = (9.0, 5.286)^T$, and is not Pareto optimal. Solving the auxiliary min–max problem (AP) with the new aspiration level, in this case, we have the new Pareto solution $(9.774, 6.226)$. Since the improvement which DM desires is not so large after solving the min–max problem (AP) with an aspiration level in many practical cases, the new aspiration level produced by automatic trade-off based on the linear approximation of Pareto surface is close to the Pareto surface. Therefore, the satisficing trade-off method using the automatic trade-off yields a desirable solution usually only in a few iterations (Fig. 2.3).

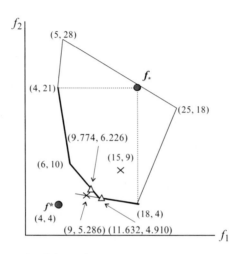

Fig. 2.3 Automatic trade-off (case 2)

Case 3. Suppose that DM wants to make f_1 less than 9.0 by all means at the moment when the first Pareto solution $(11.632, 4.910)$ is obtained. In this case, we should treat f_1 as the constraint

$$f_1(x) \leqq 9.0.$$

As will be shown in Sect. 2.3.5, the interchange between objectives and constraints can be made so easily in the formulation of auxiliary min–max problem (we can treat the criteria as DM wishes between objectives and constraints by adjusting one parameter β in the min–max problem) (Fig. 2.4).

2.3 Satisficing Trade-off Method

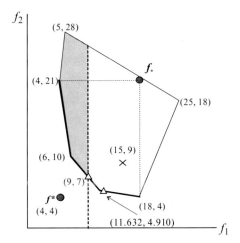

Fig. 2.4 Automatic trade-off (case 3)

2.3.4 Exact Trade-off

In linear or quadratic cases, we can evaluate the exact trade-off in an extended form of the automatic trade-off stated above. This implies that we can calculate exactly the amount of relaxation such that the new aspiration level is on the Pareto surface (Nakayama [98]). The main idea in it is that the parametric optimization technique is used in stead of the simple sensitivity analysis. Using this technique, a new Pareto solution can be obtained without solving the auxiliary scalarized optimization problem again. This implies that we can obtain a new solution very quickly. Therefore, using some graphic presentation as computer outputs, DM can see the trade-off among criteria in a dynamic way, e.g., as an animation. This makes DM's judgment easier (see Fig. 2.5).

2.3.5 Interchange Between Objectives and Constraints

In the formulation of the auxiliary scalarized optimization problem (AP), change the right-hand side of (2.5) into $\beta_i z$, namely

$$w_i(f_i(\boldsymbol{x}) - \overline{f}_i) \leqq \beta_i z. \tag{2.8}$$

As is readily seen, if $\beta_i = 1$, then the function f_i is considered to be an objective function, for which the aspiration level $\overline{f}_i$ is not necessarily attained, but the level of f_i should be better as much as possible. On the other hand, if $\beta_i = 0$, then f_i is considered to be a constraint function, for which the aspiration level $\overline{f}_i$ should be guaranteed. In many practical problems, there

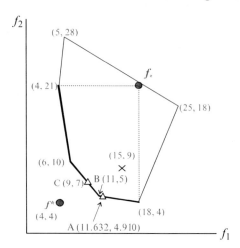

Fig. 2.5 Exact trade-off

is almost no case in which we consider the role of objective and constraint fixed from the beginning to the end, but usually we want to interchange them depending on the situation. Using (2.8), this can be done very easily (Korhonen et al. [75]). In addition, if the value of β_i is set in the middle of 0 and 1, f_i can play a role in the middle of objective and constraint which is neither a complete objective nor a complete constraint.

2.3.6 Remarks on Trade-off for Objective Functions with 0-Sensitivity

Since α is sufficiently small like 10^{-6} and $\lambda_1 + \ldots \lambda_r = 1$, we can consider in many cases

$$(\lambda_i + \alpha)w_i \simeq \lambda_i w_i.$$

When $\lambda_j = 0$ and f_j is not to be improved, we set $\Delta f_j = 0$ in the automatic trade-off. Therefore, unless we select at least one f_j with $\lambda_j \neq 0$ as an objective function to be relaxed, we cannot attain the improvement that DM wishes.

Since $(\lambda_i + \alpha)w_i$ (or approximately, $\lambda_i w_i$) can be regarded to provide the sensitivity information in the trade-off, $\lambda_i = 0$ means that the objective function f_i does not contribute to the trade-off among the objective functions. In other words, since the trade-off is the negative correlation among objective functions, $\lambda_i = 0$ means that f_i has the nonnegative correlation with some other objective functions. Therefore, if all objective functions to be relaxed, f_j, $(j \in I_R)$ have $\lambda_j = 0$ $(j \in I_R)$, then they cannot compensate for the improvement which DM wishes, because they are affected positively by some of objective functions to be improved.

2.3 Satisficing Trade-off Method

Example 2.5. Consider the following problem (Fig. 2.6):

$$\begin{aligned}
\underset{x_1, x_2, x_3}{\text{minimize}} \quad & (f_1, f_2, f_3) = (x_1, x_2, x_3)^T \\
\text{subject to} \quad & x_1 + x_2 + x_3 \geq 1, \\
& x_1 \quad\quad - x_3 \geq 0, \\
& x_1, x_2, x_3 \geq 0.
\end{aligned}$$

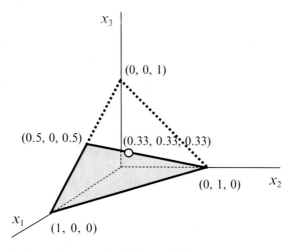

Fig. 2.6 Case of automatic trade-off with 0-sensitivity

For the first aspiration level (0.2, 0.2, 0.4), we have a Pareto value (0.333, 0.333, 0.333) and the corresponding simplex multiplier $(\lambda_1, \lambda_2, \lambda_3) = (2/3, 1/3, 0)$. Suppose that DM wants to improve f_1 and f_2, and sets their new aspiration levels 0.2 and 0.3, respectively. Since the relaxation $\Delta f_3 = 0$ by the automatic trade-off, the new aspiration level becomes (0.2, 0.3, 0.3). Associated with the new aspiration level, we have the Pareto value (0.275, 0.45, 0.275), in which neither f_2 is improved nor f_3 is relaxed. This is because that the objective functions f_1 and f_3 have a positive correlation along the edge of Pareto surface at the point (0.333, 0.333, 0.333), while f_1 and f_2 have trade-off relation with each other there. As a result, though f_3 was considered to be relaxed, it was affected strongly by f_1 and hence improved. On the other hand, despite that f_2 was considered to be improved, it was relaxed finally. This is due to the fact that the objective function to be relaxed is only f_3 despite that λ_3 is 0, and due to the fact that we did not take into account that f_3 has positive correlation with f_1. This example suggests that we should make the trade-off analysis deliberately seeing the value of simplex multiplier (Lagrange multiplier, in nonlinear cases). Like this, the satisficing trade-off method makes the DM's trade-off analysis easier by utilizing the information of sensitivity.

2.3.7 Relationship to Fuzzy Mathematical Programming

In the aspiration level approach to multiobjective programming such as the satisficing trade-off method, the wish of DM is attained by adjusting his/her aspiration level. In other words, this means that the aspiration level approach can deals with the fuzziness of right-hand side value in traditional mathematical programming as well as the total balance among the criteria. There is another method, namely *fuzzy mathematical programming*, which treats the fuzziness of right-hand side value of constraint in traditional mathematical programming. In the following, we shall discuss the relationship between the satisficing trade-off method and the fuzzy mathematical programming.

For simplicity, consider the following problem:

$$\underset{x}{\text{maximize}} \quad f_0(x) \tag{F}$$
$$\text{subject to} \quad f_1(x) = \overline{f}_1.$$

Suppose that the right-hand side value $\overline{f}_1$ is not needed to meet so strictly, but that it is fuzzy. The *membership function* for the criterion f_1 is usually given as in Fig. 2.7. Since our aim is to maximize this membership function, we can adopt the following function without change in the solution:

$$m_1(x) = \min \left\{ \frac{\overline{f}_1 - f_1(x)}{\varepsilon} + 1, \ -\frac{\overline{f}_1 - f_1(x)}{\varepsilon} + 1 \right\},$$

where ε is a parameter representing the admissible error for the target $\overline{f}_1$.

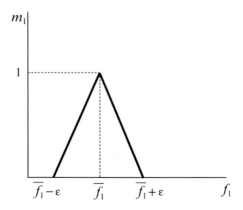

Fig. 2.7 Membership function for f_1 in (F)

2.3 Satisficing Trade-off Method

Now, the problem (F) is reduced to a kind of multiobjective optimization in which f_0 and m_1 should be maximized. Then a membership function for maximization of f_0 is usually defined with its aspiration level $\overline{f}_0$. For example,

$$m_0'(\boldsymbol{x}) = \min \left\{ -\frac{\overline{f}_0 - f_0(\boldsymbol{x})}{\overline{f}_0 - f_{0*}} + 1, \ 1 \right\}.$$

However, if we maximize the above m_0' as it is, the solution will be merely the one to the satisficing problem for which f_0 is to be just greater than $\overline{f}_0$. As was stated in the previous section, we shall use the following function in stead of m_0' in order to assure the Pareto optimality of the solution:

$$m_0(\boldsymbol{x}) = -\frac{\overline{f}_0 - f_0(\boldsymbol{x})}{\overline{f}_0 - f_{0*}} + 1.$$

Finally, our problem is to maximize both m_0 and m_1, which is usually reduced to the following problem:

$$\begin{aligned}
\underset{\boldsymbol{x},\, z}{\text{minimize}} \quad & z \\
\text{subject to} \quad & \frac{\overline{f}_0 - f_0(\boldsymbol{x})}{\overline{f}_0 - f_{0*}} - 1 \leqq z, \\
& \frac{\overline{f}_1 - f_1(\boldsymbol{x})}{\varepsilon} - 1 \leqq z, \\
& -\frac{\overline{f}_1 - f_1(\boldsymbol{x})}{\varepsilon} - 1 \leqq z.
\end{aligned}$$

Now, one may see some similarity between the above formulation and the one in the satisficing trade-off method. In the satisficing trade-off method, the objective function with target such as $f_1 \to \overline{f}_1$ is usually treated as two objective functions, $f_1 \to \max$ and $f_1 \to \min$. Under this circumstance, suppose that for $f_1 \to \max$ we set the ideal value $f_1^* = \overline{f}_1$, the nadir value $f_{1*} = \overline{f}_1 - \varepsilon$ and the aspiration level $\overline{f}_1 - \varepsilon$, and for $f_1 \to \min$ we set the ideal value $f_1^* = \overline{f}_1$, the nadir value $f_{1*} = \overline{f}_1 + \varepsilon$ and the aspiration level $\overline{f}_1 + \varepsilon$. Then the treatment of f_1 is the same between the above formulation and the satisficing trade-off method.

However, usually in the satisficing trade-off method we do not contain ε in the denominator of constraints in the min–max problem, because we make the trade-off analysis by adjusting ε rather than the target $\overline{f}_1$; for example, using (2.4) the constraint for f_1 in the min–max problem is given by

$$\frac{\overline{f}_1 - \varepsilon - f_1(x)}{f_1^* - f_{1*}} \leqq z,$$

$$-\frac{\overline{f}_1 + \varepsilon - f_1(x)}{f_1^* - f_{1*}} \leqq z.$$

With this formulation, even if DM wants $\varepsilon = 0$ and if there is no solution to $f_1(\boldsymbol{x}) = \overline{f}_1$, we can get a solution approximate to $f_1(\boldsymbol{x}) = \overline{f}_1$ as much as possible. In the fuzzy mathematical programming, however, if $\varepsilon = 0$, then we have a crisp constraint $f_1(\boldsymbol{x}) = \overline{f}_1$, and we sometimes have no feasible solution to it.

Finally as a result, we can see that the satisficing trade-off method deals with the fuzziness of right-hand side value of constraint automatically and can effectively treat problems for which fuzzy mathematical programming provides no solution. Due to this reason, we can conclude that it is better to formulate the given problem as a multiobjective optimization from the beginning and to solve it by the aspiration level approach such as the satisficing trade-off method.

2.4 Applications

Interactive multiobjective programming methods have been applied to a wide range of practical problems. Good examples in engineering applications can be seen in Eschenauer et al. [39]. One of the authors himself also has applied to several real problems: feed formulation for live stock [91,104], plastic materials blending [103], cement production [97], bond portfolio [96], erection management of cable-stayed bridges [46,105,109], scheduling of string selection in steel manufacturing [152].

In the following, some of examples are introduced briefly.

2.4.1 Feed Formulation for Live Stock

Stock farms in Japan are modernized recently. Above all, the feeding system in some farms is fully controlled by computer: Each cow has its own place to eat which has a locked gate. And each cow has a key on her neck, which can open the corresponding gate only. Everyday, on the basis of ingredient analysis of milk and/or of the growth situation of cow, the appropriate blending ratio of materials from several viewpoints should be made.

There are about 20–30 kinds of raw materials for feed in cow farms such as corn, cereals, fish meal, etc. About ten criteria are usually taken into account for feed formulation of cow:

- Cost
- Nutrition
  - Protein
  - TDN
  - Cellulose

2.4 Applications   35

- Calcium
- Magnesium
- etc.

- Stock amount of materials
- etc.

This feeding problem is well known as the diet problem from the beginning of the history of mathematical programming, which can be formulated as the traditional linear programming. In the traditional mathematical programming, the solution often occurs on the boundary of constraints. In many cases, however, the right-hand side value of constraint such as nutrition needs not to be satisfied rigidly. Rather, it seems to be natural to consider that a criterion such as nutrition is an objective function whose target has some allowable range. As was seen in the previous section, the satisficing trade-off method deals well with the fuzziness of target of such an objective function. The author and others have developed a software for feed formulation using the satisficing trade-off method, called F-STOM (feed formulation by satisficing trade-off method [91,98]). This software is being distributed to live stock farmers and feed companies in Japan through an association of live stock systems.

2.4.2 Erection Management of Cable-Stayed Bridge

In erection of cable-stayed bridge, the following criteria are considered for accuracy control [46, 105, 109]:

1. Residual error in each cable tension
2. Residual error in camber at each node
3. Amount of shim adjustment for each cable
4. Number of cables to be adjusted

Since the change of cable rigidity is small enough to be neglected with respect to shim adjustment, both the residual error in each cable tension and that in each camber are linear functions of amount of shim adjustment. Let us define n as the number of cable in use, ΔT_i $(i = 1, \ldots, n)$ as the difference between the designed tension values and the measured ones, and x_{ik} as the tension change of ith cable caused from the change of the kth cable length by a unit. The residual error in cable tension caused by the shim adjustment of $\Delta l_1, \ldots, \Delta l_n$ is given by

$$p_i = \left| \Delta T_i - \sum_{k=1}^{n} x_{ik} \Delta l_k \right|, \quad i = 1, \ldots, n.$$

Let m be the number of nodes, Δz_j ($j = 1, \ldots, m$) the difference between the designed camber values and the measured ones, and y_{jk} the camber change at jth node caused from the change of the kth cable length by a unit. Then the residual error in the camber caused by the shim adjustments of $\Delta l_1, \ldots, \Delta l_n$ is given by

$$q_j = \left| \Delta Z_j - \sum_{k=1}^{n} y_{jk} \Delta l_k \right|, \quad j = 1, \ldots, m.$$

In addition, the amount of shim adjustment can be treated as objective functions of

$$r_i = |\Delta l_i|, \quad i = 1, \ldots, n.$$

And the upper and lower bounds of shim adjustment inherent in the structure of the cable anchorage are as follows:

$$\Delta l_{Li} \leq \Delta l_i \leq \Delta l_{Ui}, \quad i = 1, \ldots, n.$$

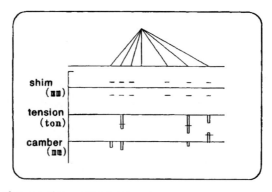

Fig. 2.8 Phase of an erection management system

Figure 2.8 shows one phase of erection management system of cable-stayed bridge using the satisficing trade-off method. The residual error of each criterion and the amount of shim adjustment are represented by graphs. The aspiration level is inputted by a mouse on the graph. After solving the auxiliary min–max problem, the Pareto solution according to the aspiration level is represented by a graph in a similar fashion. This procedure is continued until the designer can obtain a desirable shim adjustment. This operation is very easy for the designer, and the visual information on trade-off among criteria is user-friendly. The software was used for real bridge construction, say, Tokiwa Great Bridge (Ube City) and Karasuo Harp Bridge (Kita-Kyusyu City) in 1992.

2.4 Applications

2.4.3 An Interactive Support System for Bond Trading

In portfolio selection problems, many companies are now widely trying to use mathematical analysis for bond trading. In this section, some bond trading problem is formulated as a kind of multiobjective fractional problem. It will be seen in the following that the satisficing trade-off method can be effectively applied to such a portfolio problem.

Bond traders are facing almost everyday a problem which bonds and what amount they should sell and/or buy in order to attain their customers' desires. The economic environment is changing day by day, and sometimes gives us a drastic change. Bond traders have to take into account many factors, and make their decisions very rapidly and flexibly according to these changes. The number of bonds to be considered is usually more than 500, and that of criteria is about ten as will be shown later. The amount of trading is usually more than 100 million yen, and hence even a slight difference of portfolio combination might cause a big difference to profit or loss. This situation requires some effective method which helps bond traders following faithfully their value judgment on a real-time basis not only mathematically but also in such a way that their intuition fostered by their experiences can be taken in.

Bond portfolio problems are a kind of blending problems. Therefore, mathematical programming approach can be used very effectively. However, the traditional mathematical programming approach with a single objective function cannot take in the value judgment and intuition of bond traders so easily in a flexible manner for the changes of environment. We shall show that the satisficing trade-off method fits to this purpose.

Mathematical Formulation for Bond Portfolio Problems

We shall follow the mathematical model given by Konno and Inori [73]. Assume that an investor holds u_j units of bonds B_j, $j = 1, \ldots, N$. Associated with B_j, we have the following indices:

- c_j: coupon to be paid at a fixed rate (yen/bond/year)
- f_j: principal value to be refunded at maturity (yen/bond)
- p_j: present price in the market (yen/bond)
- t_j: maturity (number of years until its principal value is refunded)

Returns from bonds are the income from coupon and the capital gain due to price increase. Bond portfolio problems are to determine which bonds and what amount the investor should sell and/or buy taking into account many factors, say, expected returns and risk, the time needs money for another investment, and so on. Therefore, we have the following criteria for each bond B_j:

I. Returns

   1. Direct yield (short-term index of return)

$$\gamma_j = \frac{c_j}{p_j}$$

   2. Effective yield

$$\nu_j = \frac{[c_j\{(1+\alpha)^{t_j} - 1\}/\alpha + f_j]^{1/t_j}}{p_j} - 1,$$

   where α is the interest rate.

II. Risk

   3. Price variation

$$\pi_j = \frac{t_j}{1 + \nu_j t_j}, \quad j = 1, \ldots, N$$

Let $x_j, \ j = 1, \ldots, n_1$ and $X_k, \ k = 1, \ldots, n_2$ denote the number of bonds to be sold and to be purchased, respectively. Then S_0 and S_1 represent, respectively, the total quantity of bonds and the total value of bonds after the transaction. Namely,

$$S_0 = \sum_{j=1}^{N} u_j - \sum_{j=1}^{n_1} x_j + \sum_{k=1}^{n_2} X_k,$$

$$S_1 = \sum_{j=1}^{N} p_j u_j - \sum_{j=1}^{n_1} p_j x_j + \sum_{k=1}^{n_2} P_k X_k.$$

In addition, we set

$$S_2 = \sum_{j=1}^{N} p_j t_j u_j - \sum_{j=1}^{n_1} p_j t_j x_j + \sum_{k=1}^{n_2} P_k T_k X_k.$$

Then the average for each index is taken as an objective function:

   1'. Average direct yield

$$F_1 = \frac{\displaystyle\sum_{j=1}^{N} \gamma_j p_j u_j - \sum_{j=1}^{n_1} \gamma_j p_j x_j + \sum_{k=1}^{n_2} \gamma_k P_k X_k}{S_1}$$

2.4 Applications

2'. Average effective yield

$$F_2 = \frac{\sum_{j=1}^{N} \nu_j p_j t_j u_j - \sum_{j=1}^{n_1} \nu_j p_j t_j x_j + \sum_{k=1}^{n_2} \nu_k P_k T_k X_k}{S_2}$$

3'. Average price variation

$$F_3 = \frac{\sum_{j=1}^{N} \pi_j u_j - \sum_{j=1}^{n_1} \pi_j x_j + \sum_{k=1}^{n_2} \pi_k X_k}{S_0}$$

Our constraints are divided into soft constraints and hard constraints:

4. Average unit price

$$F_4 = \frac{\sum_{j=1}^{N} p_j u_j - \sum_{j=1}^{n_1} p_j x_j + \sum_{k=1}^{n_2} P_k X_k}{S_0}$$

5. Average maturity

$$F_5 = \frac{\sum_{j=1}^{N} t_j u_j - \sum_{j=1}^{n_1} t_j x_j + \sum_{k=1}^{n_2} T_k X_k}{S_2}$$

6. Specification of time of coupon to be paid

$$F_6 = \frac{\sum_{i \in I_1} x_i}{S_0}, \quad F_7 = \frac{\sum_{i \in I_2} x_i}{S_0},$$

where I_m is the set of indices of bonds whose coupon are paid at the time t_m.

III. Hard constraints

7. Budget constraints

$$\sum_{j=1}^{n_1} -p_j x_j + \sum_{k=1}^{n_2} P_k X_k \leq C$$

$$\sum_{k=1}^{n_2} P_k X_k \geq C$$

8. Specification of bond

$$l_j \leqq x_j \leqq u_j, \ j = 1, \ldots, n_1$$
$$L_k \leqq X_k \leqq U_k, \ k = 1, \ldots, n_2$$

For this kind of problem, the satisficing trade-off method can be effectively applied. Then we have to solve a linear fractional min–max problem. In the following section, we shall give a brief review of the method for solving it.

An Algorithm for Linear Fractional Min–Max Problems

Let each objective function in our bond trading problem be of the form $F_i(\boldsymbol{x}) = p_i(\boldsymbol{x})/q_i(\boldsymbol{x})$, $i = 1, \ldots, r$ where p_i and q_i are linear in $\boldsymbol{x}$. Then since

$$F_i^* - F_i(\boldsymbol{x}) = \frac{F_i^* q_i(\boldsymbol{x}) - p_i(\boldsymbol{x})}{q_i(\boldsymbol{x})} := \frac{f_i(\boldsymbol{x})}{g_i(\boldsymbol{x})},$$

the auxiliary min–max problem (2.2) becomes a kind of linear fractional min–max problem. For this kind of problem, several methods have been developed: Here we introduce a Dinkelbach-type algorithm (Borde–Crouzeix [11] and Ferland–Potvin [42]) as is stated in the following:

Step 1. Let $\boldsymbol{x}^0 \in X$. Set $\theta^0 = \max_{1 \leq i \leq r} f_i(\boldsymbol{x}^0)/g_i(\boldsymbol{x}^0)$ and $k = 0$.

Step 2. Solve the problem

$$T_k(\theta^k) = \min_{\boldsymbol{x} \in X} \max_{1 \leq i \leq r} \frac{f_i(\boldsymbol{x}) - \theta^k g_i(\boldsymbol{x})}{g_i(\boldsymbol{x}^k)}. \qquad (P_k)$$

Let $\boldsymbol{x}^{k+1}$ be a solution to (P_k).

Step 3. If $T_k(\theta^k) = 0$ then stop: θ^k is the optimal value of the given min–max Problem, and $\boldsymbol{x}^{k+1}$ is the optimal solution.

If not, take $\theta^{k+1} = \max_{1 \leq i \leq r} f_i(\boldsymbol{x}^{k+1})/g_i(\boldsymbol{x}^{k+1})$. Replace k by $k + 1$ and go to Step 2.

Note that the problem (P_k) is the usual linear min–max problem. Therefore, we can obtain its solution by solving the following equivalent problem in a usual manner:

$$\begin{aligned} &\underset{\boldsymbol{x}, z}{\text{minimize}} && z && (Q_k) \\ &\text{subject to} && \frac{f_i(\boldsymbol{x}) - \theta^k g_i(\boldsymbol{x})}{g_i(\boldsymbol{x}^k)} \leqq z, \ i = 1, \ldots, r. \end{aligned}$$

2.4 Applications

An Experimental Result

A result of our experiments is shown below: The problem is to decide a bond portfolio among 37 bonds selected from the market. The holding bonds are $x(1) = 5,000$, $x(9) = 1,000$, $x(13) = 2,500$, $x(17) = 4,500$, $x(19) = 5,500$, $x(21) = 6,000$, $x(23) = 5,200$, $x(25) = 4,200$, $x(27) = 3,200$ and $x(37) = 3,800$. The experiment was performed by a worker of a security company in Japan who has a career of acting as a bond trader.

	Pareto sol.	Asp. level (target range)	Lowest	Highest	Sensitivity
F1 (max)	5.8335	5.8000	5.4788	5.9655	0.0133
F2 (max)	6.8482	6.8000	6.7165	6.8698	0.0000
F3 (min)	0.1292	0.1300	0.1262	0.1325	1.0000
* F4	102.7157	$F4 \leq 103.00$			
* F5		$4.0000\ 4.00 \leq F5 \leq\ 5.00$			
* F6		$0.2000\ 0.20 \leq F6$			
* F7		$0.2000\ 0.20 \leq F7$			

$$
\begin{array}{ll}
x(\ 1) = 1{,}042.0643 & x(\ 2) = 400.0000 \\
x(\ 3) =\ \ 400.0000 & x(\ 4) = 200.0000 \\
x(\ 5) =\ \ \ \ \ 0.0000 & x(\ 6) = 400.0000 \\
x(\ 7) =\ \ \ \ \ 0.0000 & x(\ 8) =\ \ \ \ 0.0000 \\
x(\ 9) =\ \ 547.3548 & x(10) =\ \ \ \ 0.0000 \\
x(11) =\ \ \ \ \ 0.0000 & x(12) =\ \ \ \ 0.0000 \\
x(13) = 2{,}500.0000 & x(14) = 200.0000 \\
x(15) =\ \ \ \ \ 0.0000 & x(16) = 200.0000 \\
x(17) = 4{,}500.0000 & x(18) =\ \ \ \ 0.0000 \\
x(19) = 5{,}321.9573 & x(20) =\ \ \ \ 0.0000 \\
x(21) = 6{,}000.0000 & x(22) =\ \ \ \ 0.0000 \\
x(23) = 5{,}200.0000 & x(24) =\ \ \ \ 0.0000 \\
x(25) = 4{,}200.0000 & x(26) =\ \ \ \ 0.0000 \\
x(27) = 3{,}200.0000 & x(28) = 274.6025 \\
x(29) =\ \ 400.0000 & x(30) = 200.0000 \\
x(31) =\ \ \ \ \ 0.0000 & x(32) = 200.0000 \\
x(33) =\ \ 400.0000 & x(34) = 400.0000 \\
x(35) =\ \ \ \ \ 0.0000 & x(36) =\ \ \ \ 0.0000 \\
x(37) = 3{,}800.0000 &
\end{array}
$$

The asterisk of F4–F7 implies soft constraints. In this system, we can change objective function into soft constraints and vice versa. Here, the bond trader changed F2 (effective yield) into a soft constraint, and F4 (unit price) into an objective function. Then under the modified aspiration level by trade-off, the obtained result is as follows:

	Pareto sol.	Asp. level (target range)	Lowest	Highest	Sensitivity
F1 (max)	5.8629	5.9000	5.8193	5.9608	0.0000
* F2 (max)	6.8482	6.8482 ≤ F2			
F3 (min)	0.1302	0.1292	0.1291	0.1322	1.0000
F4 (min)	102.3555	102.1000	102.0676	102.7228	0.0043
* F5		4.0000 4.00 ≤ F5 ≤	5.00		
* F6		0.2000 0.20 ≤ F6			
* F7		0.2000 0.20 ≤ F7			

$$
\begin{array}{ll}
x(\ 1)= & 0.0000 \qquad x(\ 2)= 400.0000 \\
x(\ 3)= & 400.0000 \qquad x(\ 4)= 200.0000 \\
x(\ 5)= & 0.0000 \qquad x(\ 6)= 400.0000 \\
x(\ 7)= & 0.0000 \qquad x(\ 8)= \ \ 0.0000 \\
x(\ 9)= & 139.4913 \qquad x(10)= \ \ 0.0000 \\
x(11)= & 0.0000 \qquad x(12)= \ \ 0.0000 \\
x(13)= & 2{,}500.0000 \qquad x(14)= 381.5260 \\
x(15)= & 0.0000 \qquad x(16)= 200.0000 \\
x(17)= & 4{,}500.0000 \qquad x(18)= \ \ 0.0000 \\
x(19)= & 5{,}277.1577 \qquad x(20)= \ \ 0.0000 \\
x(21)= & 6{,}000.0000 \qquad x(22)= \ \ 0.0000 \\
x(23)= & 5{,}200.0000 \qquad x(24)= \ \ 0.0000 \\
x(25)= & 4{,}200.0000 \qquad x(26)= \ \ 0.0000 \\
x(27)= & 3{,}200.0000 \qquad x(28)= \ 14.8920 \\
x(29)= & 400.0000 \qquad x(30)= 200.0000 \\
x(31)= & 0.0000 \qquad x(32)= 200.0000 \\
x(33)= & 400.0000 \qquad x(34)= 400.0000 \\
x(35)= & 400.0000 \qquad x(36)= 222.7743 \\
x(37)= & 3{,}800.0000
\end{array}
$$

2.5 Some Remarks on Applications

Decision making is a problem of value judgment. One of most important tasks in multiobjective optimization is how to treat the value judgment of decision maker (DM). In order to get a solution reflecting faithfully the value judgment of DM in a flexible manner for the multiplicity of value judgment and complex changes of environment of decision making, cooperative systems of man and computers are very attractive: above all, interactive multiobjective programming methods seem promising.

Among several interactive multiobjective programming techniques, the aspiration level approach has been applied to several kinds of real problems, because:

2.5 Some Remarks on Applications

1. It does not require any consistency of judgment of DM.
2. It reflects the value of DM very well.
3. It is easy to implement.

In particular, the point (1) is very important, because DM tends to change his attitude even during the decision making process. This implies that the aspiration level approach such as the satisficing trade-off method can work well not only for the multiplicity of value judgment of DMs but also for the dynamics of value judgment of them.

Even in cases with a large number of objective functions, the aspiration level approach works well, because aspiration levels are intuitive for DM and reflect the value of DM very well. Using graphic user interface, DM can input the aspiration level very easily, and decrease the burden by using automatic (or, exact) trade-off method as was stated in previous section.

Trade-off analysis is relatively easily made if we know the whole configuration of Pareto frontier. At least in a neighborhood of the obtained Pareto solution, the information on the configuration of Pareto frontier is needed for trade-off analysis. For many smooth cases under some appropriate conditions, Lagrange multipliers for auxiliary scalarized problems provide those information. However, we cannot utilize Lagrange multiplier as trade-off information in nonsmooth cases such as discrete problems. Evolutionary multiobjective optimization (EMO) methods, which have been studied actively in recent years, work well to generate Pareto frontiers not only in smooth cases but also nonsmooth (even discrete) cases. However, it is in general time consuming to generate the whole of Pareto frontier. In many practical problems, DM is interested in some part of Pareto frontier, but not in the whole. Therefore, it is important to combine the aspiration level approach and EMO methods effectively. This will be discussed in subsequent chapters.

Chapter 3
Generation of Pareto Frontier by Genetic Algorithms

Recently, multiobjective optimization methods using evolutionary methods such as *genetic algorithms* have been studied actively by many researchers [4, 23, 24, 30, 33, 37, 44, 49, 59, 111–113, 131, 149, 160, 161, 169]. These approaches are useful for generating Pareto frontiers in particular with two or three objective functions, and decision making can be easily performed on the basis of visualized Pareto frontier. In generating Pareto frontiers by evolutionary methods, there are two main issues of the *convergence* and the *diversity*: (1) how to guide individuals to the real Pareto frontier as close and fast as possible and (2) how to keep the diversity of individuals spreading over the whole of Pareto frontier at the final generation. The convergence and the diversity of individuals are closely related to *fitness evaluation* for each individual and *density estimation* in a population. In this chapter, we describe several techniques for fitness evaluation and density estimation, and introduce representative algorithms using genetic algorithm.

3.1 Evolutionary Multiobjective Optimization

Genetic algorithms (GA) were developed by Holland [57, 58], and later were applied to problems on optimization of function by De Jong [29]. After publishing the book "Genetic Algorithms in Search, Optimization and Machine Learning" by Goldberg [49], GAs have been paid considerable attention as a useful tool of optimization.

Here, the simple procedure of GA based on the *natural selection* and the *natural genetics* can be summarized as follows (see Fig. 3.1), and more details on GA operators can be referred to the book [30]:

Step 1. (Initialization)   Generate randomly an initial population with a given size.

Step 2. (Evaluation)   Assign a fitness for each individual in the population.

H. Nakayama et al., *Sequential Approximate Multiobjective Optimization*
Using Computational Intelligence, Vector Optimization,
DOI 10.1007/978-3-540-88910-6_3, © 2009 Springer-Verlag Berlin Heidelberg

Step 3. (Selection) Select a new population on the basis of the assigned fitness.

Step 4. (Termination) Terminate the iteration if a stop condition, e.g., the number of generations, holds. Otherwise go to the next step.

Step 5. (Crossover) Make pairs randomly and a crossover for each pair according to a given crossover rate (probability).

Step 6. (Mutation) Mutate each individual according to a given mutation rate (probability).

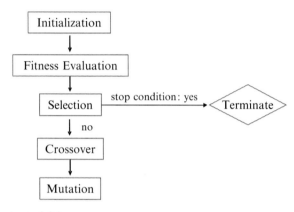

Fig. 3.1 Flowchart of GA

GA is a kind of search algorithms using *population-based approach*, thus is suited to generate exactly or approximately the set of Pareto optimal solutions to multiobjective optimization problem:

$$\underset{x}{\text{minimize}} \quad \boldsymbol{f}(\boldsymbol{x}) = (f_1(\boldsymbol{x}), \ldots, f_m(\boldsymbol{x}))^T \quad \text{(MOP)}$$
$$\text{subject to} \quad \boldsymbol{x} \in X \subset \mathbb{R}^n,$$

where $\boldsymbol{x}$ is a design variable and X is a set of feasible solutions (or constraint set).

3.1.1 Vector Evaluated Genetic Algorithm

As an extension of a single objective GA for multiobjective optimization problem, Schaffer [131] first proposed the *vector evaluated genetic algorithm* (VEGA), in which subpopulations of the next generation are reproduced from the current population according to each of the objective functions, separately. Figure 3.2 illustrates the schematic of VEGA.

3.1 Evolutionary Multiobjective Optimization

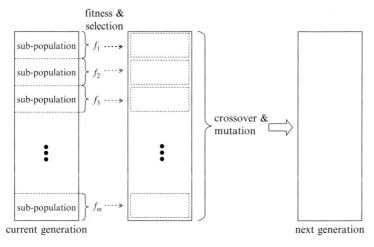

Fig. 3.2 Schematic of VEGA

VEGA may be the simplest among multiobjective GAs. However since VEGA assigns the fitness to each individual in a subpopulation with one of the objective functions, it has a tendency to generate individuals in the vicinity of the solutions which minimize each objective function.

3.1.2 Multiobjective Genetic Algorithm

On the other hand, Fonseca–Flemming [44] proposed the *multiobjective genetic algorithm* (MOGA). MOGA uses the *ranking method*[1] in the fitness assignment of each individual.

Consider an individual x^o which is dominated by n_o individuals in the current population, then its *rank* r^o is given by

$$r^o = n_o + 1. \tag{3.1}$$

From the above, it can be known that all nondominated (Pareto optimal) individuals are assigned the rank = 1. The ranking method based on the Pareto optimality among individuals has a merit to be computationally simple.

Originally, the concept of *rank* was introduced by Goldberg [49]. In the ranking method of Goldberg, all nondominated individuals are assigned the rank = 1 first. Then the individuals with rank = 1 are removed from the population. All nondominated individuals in the remained individuals are assigned

---

[1] In this book, the fitness evaluation based on the Pareto optimality or Pareto ordering is called the *ranking method* or *rank-based method*.

the rank = 2. This procedure is repeated until the ranks are assigned to all individuals. This ranking method is followed by the *nondominated sorting genetic algorithm* (NSGA) by Srinibas–Deb [144]. Figure 3.3 shows the difference between these two ranking methods.

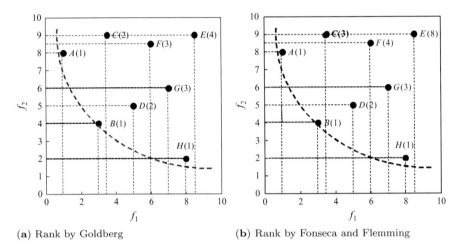

(a) Rank by Goldberg   **(b)** Rank by Fonseca and Flemming

Fig. 3.3 Ranking methods: each number in *parentheses* represents the rank of each individual and the *curve* represents the true Pareto frontier

Another feature of MOGA is the fitness assignment introducing the *niching* in the objective function space among individuals with the same rank, in order to keep the diversity among nondominated individuals. The procedure of the fitness assignment by MOGA is as follows:

Step 1. Calculate the rank r^i for all individuals x^i, $i = 1, \ldots, N$ by (3.1), where N is a population size.

Step 2. For each individual x^i, the *average fitness* F_i is assigned by

$$F_i = \begin{cases} N - \sum_{k=1}^{r^i-1} \mu(k) - 0.5\left(\mu(r^i) - 1\right) & \text{if } r^i \geqq 2 \\ N - 0.5\left(\mu(1) - 1\right) & \text{if } r^i = 1 \end{cases},$$

where $\mu(k)$ represents the number of individuals with the rank k.

Step 3. Calculate a *niche count* nc_i for an individual x^i with the rank r:

$$nc_i = \sum_{j=1}^{\mu(r)} Sh(d_{ij}), \quad i = 1, \ldots, \mu(r),$$

3.1 Evolutionary Multiobjective Optimization

where $Sh(d_{ij})$ is the *sharing function* (Fig. 3.4):

$$Sh(d_{ij}) = \begin{cases} 1 - \dfrac{d_{ij}}{\sigma_{share}} & \text{if } d_{ij} \leqq \sigma_{share} \\ 0 & \text{otherwise} \end{cases},$$

- σ_{share} is a fixed parameter of *sharing radius*.
- $d_{ij} = \sqrt{\sum_{k=1}^{m} \left(\dfrac{f_k^i - f_k^j}{f_k^{max} - f_k^{min}} \right)^2}$ for two individuals $\boldsymbol{x}^i$ and $\boldsymbol{x}^j$.
- f_k^{max} and f_k^{min} are the maximal and minimal value of kth objective function, respectively.

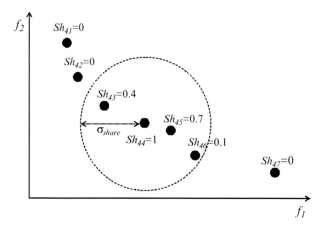

Fig. 3.4 Sharing function and niche count: for example, $nc_4 = \sum_{j=1}^{7} Sh_{4j} = 0 + 0 + 0.4 + 1 + 0.7 + 0.1 + 0 = 2.2$, where $Sh(d_{4j})$ is denoted by Sh_{4j} for simplicity

Next for an individual $\boldsymbol{x}^i$ with the rank r, the *shared fitness* is calculated by

$$F_i' = \frac{F_i}{nc_i}, \quad i = 1, \ldots, \mu(r),$$

and in order to preserve the same average fitness, the shared fitness is scaled by

$$F_i' \leftarrow \frac{F_i \mu(r)}{\sum_{j=1}^{\mu(r)} F_j'} F_i', \quad i = 1, \ldots, \mu(r),$$

which is called the *scaled fitness*. Starting with a rank $r = 1$ and renewing by $r = r + 1$, this step is repeated for all possible ranks.

Note that MOGA may assign worse fitness to $\boldsymbol{x}^1$ than $\boldsymbol{x}^2$ even though the rank for $\boldsymbol{x}^1$ is better than the one for $\boldsymbol{x}^2$ depending on the density (distribution) of population or the sharing parameter σ_{share}. Deb claimed that this phenomenon leads to a slow convergence or inability to find a good spread in the Pareto frontier [30]. Figure 3.5 and Table 3.1 show the fitness assignment for individuals with the rank = 1 and the rank = 2, supposing the population size $N = 20$ and the sharing radius $\sigma_{share} = 0.5$. As seen from these figure and table, although the individuals B and C have the rank = 1, their scaled fitness are worse than the individuals E and G with the rank = 2.

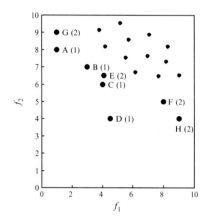

Fig. 3.5 Data of Table 3.1

Table 3.1 Fitness assignment of MOGA

	f_1	f_2	Rank	Average fitness	Niche count	Shared fitness	Scaled fitness
A	1	8	1	18.5	1.832	10.100	21.964
B	3	7	1	18.5	2.599	7.118	15.478
C	4	6	1	18.5	2.584	7.160	15.571
D	4.5	4	1	18.5	1.917	9.651	20.987
E	4.1	6.5	2	14.5	1.368	10.601	15.847
F	8	5	2	14.5	1.881	7.707	11.520
G	1	9	2	14.5	1.204	12.048	18.010
H	9	4	2	14.5	1.717	8.444	12.623
⋮					⋮		

Next, we introduce two methods overcoming the difficulty stated above, NSGA-II [32, 33] and SPEA2 [170], which are most commonly used as *elitist multiobjective evolutionary algorithms*.

3.1.3 Elitist Nondominated Sorting Genetic Algorithm (NSGA-II)

Before explaining NSGA-II, we first introduce NSGA proposed by Srinibas–Deb [144]. Main features of NSGA are to use:

- A nondominated sorting based on the ranking method by Goldberg [49]
- A fitness assignment in order to regard nondominated individuals as more important
- A sharing strategy in the design space in order to preserve the diversity among solutions in the Pareto frontier

However, it has been observed that the performance of NSGA depends sensitively on parameters for the sharing strategy.

Later, Deb et al. [32] suggested NSGA-II with the following characteristics:

- To use a *crowded tournament selection*
- To use a *crowding distance* without using any parameters such as the sharing parameter
- To preserve nondominated individuals which were found at intermediate generations

The algorithm of NSGA-II is summarized as follows (Fig. 3.6):

Step 1. Combining a parent population P_t and an offspring population Q_t, generate $R_t = P_t \cup Q_t$.

Step 2. Calculate the ranks of individuals in a population R_t by the ranking method of Goldberg, which is a *nondominated sorting* of NSGA-II.

Step 3. Classify all individuals according to their ranks, and let F_i be a subpopulation of individuals with the same rank i, $i = 1, 2, \ldots$.

Step 4. Generate a new population $P_{t+1} = \emptyset$, and let $i = 1$. Renew $P_{t+1} = P_{t+1} \cup F_i$ and $i = i + 1$ until $|P_{t+1}| + |F_i| < N$, where N is a population size and $|\cdot|$ represents the number of individuals in a population.

Step 5. By using the *crowding distance*, select $N - |P_{t+1}|$ individuals with a large crowding distance from F_i, and add to P_{t+1}.

Crowding distance:

- Generate a subpopulation F_i of individuals with the same rank.
- Sort F_i according to each objective function, and the sorted F_i is denoted by F'_{ik} for kth objective function, $k = 1, \ldots, m$.
- For the jth individual $\boldsymbol{x}_j$ in the sorted subpopulation F'_{ik}, calculate the distance d_{jk} between the values of the kth objective function for the $(j-1)$th individual $\boldsymbol{x}_{(j-1)}$ and the $(j+1)$th individual $\boldsymbol{x}_{(j+1)}$. Here, set the distance $d_{jk} = \infty$ for an individual $\boldsymbol{x}_j$ with the maximal or minimal value for the kth objective function.

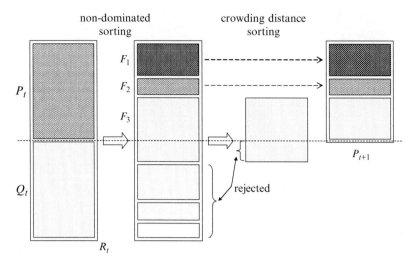

Fig. 3.6 Schematic of NSGA-II

- Add the distances d_{jk}, $k = 1, \ldots, m$ for all objective functions:

$$d_j = \sum_{k=1}^{m} \frac{d_{jk}}{f_k^{max} - f_k^{min}},$$

which is the *crowding distance* of an individual x_j so as to consider the density of individuals distributed near x_j (Fig. 3.7).

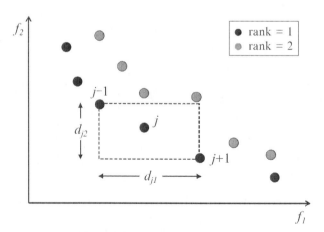

Fig. 3.7 Crowding distance in NSGA-II: for example, the crowding distance of x_j in the objective function space is given by $d_{j1} + d_{j2}$

3.1 Evolutionary Multiobjective Optimization

Step 6. Generate an offspring population Q_{t+1} from P_{t+1} by using the crowded tournament selection, crossover and mutation. The *crowded tournament selection operator* means that an individual x_i is preferred to another individual x_j if either of the following conditions holds:

> **Crowded tournament selection operator:**
> - x_i has a smaller rank than x_j.
> - x_i has a larger crowding distance than x_j when the ranks of x_i and x_j are the same.

NSGA-II introduces the crowding distance in order to maintain the diversity among nondominated individuals. As a result, NSGA-II has been observed to provide well-distributed nondominated individuals in the objective function space. However, at latter generations, when nondominated individuals (i.e., with the rank $= 1$) are more than N numbers in the population combining parent and offspring populations, a part of them are deleted by the crowded tournament selection operator. The deleted individuals may have some important information on generating the Pareto frontier, and because of this, NSGA-II may lose its convergence to the real Pareto frontier.

3.1.4 Strength Pareto Evolutionary Algorithm (SPEA2)

Zitzler–Thiele [167–169] suggested the so-called *strength Pareto evolutionary algorithm* (SPEA) which uses new mechanisms described below:

- A fitness assignment which considers how many individuals it dominates and it is dominated by.
- An external population called *archive* to which the nondominated individuals at each generation are saved, keeping a constant size of population. Besides, an archive is utilized for a fitness assignment in the current population.
- A clustering method in order to preserve the diversity of nondominated individuals in the objective function space. When the number of nondominated individuals exceeds given a population size, some of them are removed by some clustering method.

Later, Zitzler et al. [170] suggested SPEA2 improving SPEA in terms of a new fitness assignment and *archive truncation method*. The procedure of SPEA2 is summarized as follows:

Step 1. Initialize the parameters.

- N: population size, $\overline{N}$: archive size (external population size)
- T: maximal number of generations, $t = 0$: initial generation
- P_0: initial population generated randomly, $\overline{P}_0 = \emptyset$: initial archive

Step 2. Calculate the fitness of each individual in the population combining the current population P_t and archive $\overline{P}_t$.

Fitness assignment: for an individual x_i,

- Calculate the so-called *raw fitness* $R(i)$ which means the *strengths of dominators* of an individual (Fig. 3.8):
  - If x_i is nondominated, the fitness is assigned by $R(i) = 0$.
  - If x_i is dominated by another individuals x_j, $j = 1, \ldots, k$,
    · Count the number of individuals $S(j)$ which x_j dominates.
    · Then the fitness is assigned by $R(i) = \sum_{j=1}^{k} S(j)$.

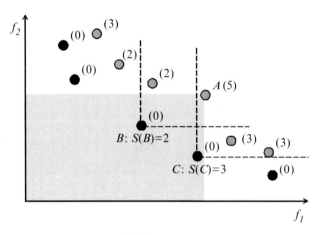

Fig. 3.8 Raw fitness assignment in SPEA2: A is dominated by B and C, where B dominates two individuals, while C three individuals. Thus the raw fitness for A is assigned by the sum of $R(A) = S(B) + S(C) = 5$

- Estimate the density $D(i)$ which is to discriminate among the individuals with the same raw fitness:
  - Calculate the distances d_i^j, $j = 1, \ldots, N + \overline{N}$ in the objective function space between x_i and all individuals.
  - Sort the distances d_i^j in increasing order, and denote the kth distance by d_i^k.
  - Calculate the density $0 < D(i) < 1$ defined by

$$D(i) = \frac{1}{d_i^k + 2},$$

where $k = \sqrt{N + \overline{N}}$ is given by the square root of the population size [141].

3.1 Evolutionary Multiobjective Optimization

- Assign finally the fitness by adding the raw fitness and the density:

$$F(i) = R(i) + D(i).$$

Step 3. Save the nondominated individuals with the raw fitness $R(i) = 0$ to $\overline{P}_{t+1}$, where there are no duplicated individuals in the design space. If $|\overline{P}_{t+1}|$ exceeds $\overline{N}$, then delete excess individuals by using the *truncation method*. Otherwise, $\overline{P}_{t+1}$ is filled to the size $\overline{N}$ with dominated individuals This step is called *environmental selection* in SPEA2:

Truncation method (Fig. 3.9):

- Find the nearest two individuals A and B in the objective function space.
- Calculate the distances d_A and d_B of the next nearest individual from A and B in the objective function space, respectively.
- Delete the individual with the smaller one between d_A and d_B.

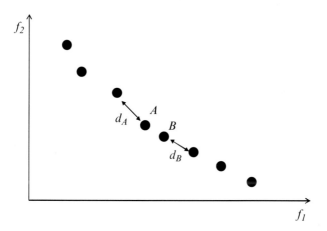

Fig. 3.9 Archive truncation method: for the nearest two points A and B, they have the next nearest distance, d_A and d_B. In this case, B is deleted since $d_A > d_B$

Step 4. Terminate the iteration if the stop condition holds, e.g., $t \geqq T$. Otherwise, go to Step 5.
Step 5. Choose N individuals from $\overline{P}_{t+1}$ by *binary tournament selection*, which is called *mating selection*.
Step 6. Generate a new population P_{t+1} by using crossover and mutation operators. Set $t = t + 1$ and go to Step 2.

Although the effectiveness of SPEA2 has been observed, SPEA2 needs long run-time in the fitness assignment and the calculation of truncation operator.

3.2 Fitness Evaluation Using DEA

3.2.1 Quantitative Evaluation of Fitness

As described above, most of fitness evaluations are based on Pareto dominance relation (*rank*) in order to guide the population to the real Pareto frontier as fast as possible. Sometimes, the fitness by the rank may be amended so as to incorporate the diversity of nondominated individuals in the objective function space. Basically, however, the rank-based fitness indicates the numbers of dominant or dominated individuals, and is of ordinal scale. Hence the conventional methods for fitness evaluation do not reflect the "distance" of each individual in the objective function space from an approximate Pareto frontier.[2]

For example, as shown in Fig. 3.10, D and F is very close, but the rank of F is worse than the rank of D. In addition, G has the same rank as F, in spite that G is further than F from the Pareto frontier. One may see that some measure reflecting the distance between each individual and Pareto frontier would provide a better convergence property. This encourages us to apply data envelopment analysis (DEA) [18, 19].

Data envelopment analysis (DEA) is a method to measure the *relative efficiency* of decision making units (DMUs) performing similar tasks in a production system that consumes multiple inputs to produce multiple outputs. DEA has the following main characteristics:

- It can be applied to analyze multiple outputs and multiple inputs without preassigned weights.
- It can be used for measuring a relative efficiency based on the observed data without knowing information on the production function.
- Decision makers' preferences can be incorporated in DEA models.

DEA was originally suggested by Charnes–Cooper–Rhodes (CCR), and built on the idea of Farrell [40] which is concerned with the estimation of technical efficiency and efficient frontiers. So far, there are *CCR model* [18, 19], *BCC model* [6] and *FDH model* [151] as representative models.

---

[2] In this chapter, nondominated individuals generated by GA is called an approximate Pareto optimal solution, and in the objective function space, an approximate Pareto optimal value or an approximate Pareto frontier.

3.2 Fitness Evaluation Using DEA

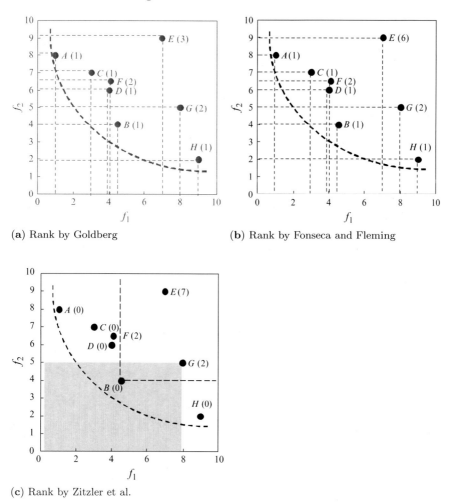

Fig. 3.10 A problem of rank-based methods: the dotted curve is a Pareto frontier, and the rank does not necessarily reflect the distance from the Pareto frontier

3.2.2 Data Envelopment Analysis

CCR model [18, 19] generalized the single output/single input ratio efficiency measure for each DMU to multiple outputs/multiple inputs situations by forming the ratio of a weighted sum of outputs to a weighted sum of inputs. Later, Banker–Charnes–Cooper (BCC) suggested a model for estimating technical efficiency and scale inefficiency in DEA. BCC model [6] relaxed the constant *returns to scale* assumption of CCR model and made it possible to investigate whether the performance of each DMU was conducted in region of increasing, constant or decreasing returns to scale in multiple outputs and

58   3 Generation of Pareto Frontier by GA

multiple inputs situations. In addition, Tulkens [151] introduced a relative efficiency to nonconvex *free disposable hull*[3] (FDH) of the observed data defined by Deprins et al. [34], and formulated a mixed integer programming to calculate the relative efficiency for each DMU. These models are classified by how to determine the *production possibility set*; a convex cone, a convex hull and a free disposable hull of observed data set. The efficient (or Pareto) frontier generated from observed data by DEA is called the *DEA-efficient frontier*, and DMU is said to be *DEA efficient* if the *DEA efficiency* (efficiency evaluated by DEA) is equal to one, which means that DMU is located on *DEA-efficient frontier*.

Here, we describe the DEA efficiency, which is the *ratio efficiency*, along a simple example.

Example 3.1. Consider the following example as shown in Table 3.2.

Table 3.2 Case of single input and single output

DMU	A	B	C	D	E	F
Input (x)	3	4	9	7	6	8
Output (y)	2	4	7	3	5	5.5
Output/input	0.667	1.000	0.778	0.429	0.833	0.688

In DEA, the efficiency of each DMU is given by the *ratio value*

$$\mathrm{R} = \frac{\text{output}}{\text{input}} = \frac{y}{x}.$$

Then, it can be known from Table 3.2 that since the ratio value of DMU B is maximal, DMU B has the best performance among six DMUs. The points on the solid line shown in Fig. 3.11 have the same ratio value with DMU B, this line is called the *DEA-efficient frontier*. A DMU is said to be *DEA-efficient* if it is on the DEA-efficient frontier, otherwise *DEA-inefficient*.

In addition, the *relative ratio value* of R_o, $o = A, \ldots, F$ to R_B is defined by

$$\theta_o = \frac{\mathrm{R}_o}{\mathrm{R}_B}, \ o = A, \ldots, F,$$

which is used as the *relative efficiency* of each DMU in DEA (see Table 3.3).

For example, comparing with DMU B, it means that DMU D ($\theta_D = 0.429$) has the productivity of 43% of DMU B per one input.

---

[3] The *free disposable hull* is the set consisting of any points that perform less output with the same amount of input as observed points, and/or those that perform more input with the same amount of output.

3.2 Fitness Evaluation Using DEA

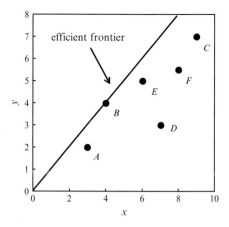

Fig. 3.11 Efficient frontier by DEA (Example 3.1): the *solid line* is the DEA-efficient frontier

Table 3.3 DEA efficiency of Example 3.1

DMU	A	B	C	D	E	F
Efficiency (θ_o)	0.667	1.000	0.778	0.429	0.833	0.688

3.2.3 DEA Models

Hereafter, the following notations are commonly used in this section:

- q, p, m: the number of DMU, output and input, respectively
- DMU_o: an object of evaluation $o = 1, \ldots, q$
- y_{kj}: kth output of DMU_j, $j = 1, \ldots, q$; $k = 1, \ldots, p$
- x_{ij}: ith input of DMU_j, $j = 1, \ldots, q$; $i = 1, \ldots, m$
- $\boldsymbol{y}_j := (y_{1j}, \ldots, y_{pj})^T$: output vector of DMU_j, $j = 1, \ldots, q$
- $\boldsymbol{x}_j := (x_{1j}, \ldots, x_{mj})^T$: input vector of DMU_j, $j = 1, \ldots, q$
- ε: sufficiently small positive number, e.g., 10^{-7}

Extending to the case with multiple inputs and multiple outputs, the relative efficiency with the input weights ν_i, $i = 1, \ldots, p$ and the output weights μ_k, $k = 1, \ldots, m$ is defined by

$$\frac{\sum_{k=1}^{p} \mu_k y_{ko}}{\sum_{i=1}^{m} \nu_i x_{io}}. \tag{3.2}$$

Since the above formula (3.2) of ratio value increases infinitely, we impose the normality condition that the value of (3.2) is not larger than one for all DMUs. Then, the following fractional linear programming problem (FLP) for DMU$_o$ can be formulated in order to determine the optimal weights $\boldsymbol{\mu}$ and $\boldsymbol{\nu}$:

$$\underset{\boldsymbol{\mu}, \boldsymbol{\nu}}{\text{maximize}} \qquad \frac{\sum_{k=1}^{p} \mu_k y_{ko}}{\sum_{i=1}^{m} \nu_i x_{io}} \qquad \text{(FLP)}$$

$$\text{subject to} \qquad \frac{\sum_{k=1}^{p} \mu_k y_{kj}}{\sum_{i=1}^{m} \nu_i x_{ij}} \leq 1, \quad j = 1, \ldots, q,$$

$$\mu_k \geq \varepsilon, \quad k = 1, \ldots, p,$$

$$\nu_i \geq \varepsilon, \quad i = 1, \ldots, m.$$

Also the problem (FLP) can be reformulated as the following equivalent linear programming problem (CCR):

$$\underset{\boldsymbol{\mu}, \boldsymbol{\nu}}{\text{maximize}} \qquad \sum_{k=1}^{p} \mu_k y_{ko} \qquad \text{(CCR)}$$

$$\text{subject to} \qquad \sum_{i=1}^{m} \nu_i x_{io} = 1,$$

$$\sum_{k=1}^{p} \mu_k y_{kj} - \sum_{i=1}^{m} \nu_i x_{ij} \leq 0, \quad j = 1, \ldots, q,$$

$$\mu_k \geq \varepsilon, \quad k = 1, \ldots, p,$$

$$\nu_i \geq \varepsilon, \quad i = 1, \ldots, m,$$

and its dual problem (CCR$_D$) is formulated by

$$\underset{\theta, \boldsymbol{\lambda}, \boldsymbol{s}_x, \boldsymbol{s}_y}{\text{minimize}} \qquad \theta - \varepsilon(\mathbf{1}^T \boldsymbol{s}_x + \mathbf{1}^T \boldsymbol{s}_y) \qquad \text{(CCR}_D\text{)}$$

$$\text{subject to} \qquad \boldsymbol{X}\boldsymbol{\lambda} - \theta \boldsymbol{x}_o + \boldsymbol{s}_x = \mathbf{0},$$

$$\boldsymbol{Y}\boldsymbol{\lambda} - \boldsymbol{y}_o - \boldsymbol{s}_y = \mathbf{0},$$

$$\boldsymbol{\lambda} \geq \mathbf{0}, \quad \boldsymbol{s}_x \geq \mathbf{0}, \quad \boldsymbol{s}_y \geq \mathbf{0}.$$

The above problem (CCR) is called CCR model [18, 19] which is the most basic model among the DEA models. By using CCR model, the relative efficiency is called the *CCR efficiency*, and the efficient frontier is called the *CCR-efficient frontier*[4] which was stated in the previous section (also see Fig. 3.11).

Definition 3.1 (CCR Efficiency or DEA Efficiency). For the optimal solution θ^*, $\boldsymbol{s}_x^*$, $\boldsymbol{s}_y^*$ to the problem (CCR), DMU$_o$ is said to be *CCR-efficient* (or *DEA-efficient*) if $\theta^* = 1$, $\boldsymbol{s}_x^* = \mathbf{0}$ and $\boldsymbol{s}_y^* = \mathbf{0}$.

---

[4] In this book, we use the terminologies of DEA efficiency and DEA-efficient frontier as the ones in CCR model.

3.2 Fitness Evaluation Using DEA

In CCR model, the *production possibility set* is the *convex cone* (or conical hull) generated by the observed data, since one takes a viewpoint of the fact that the scale efficiency of a DMU is constant, that is to say, *constant returns to scale*.

On the other hand, BCC model by Banker–Charnes–Cooper [6] is formulated similarly to that for CCR model. The dual problem for BCC model is obtained by adding the convexity constraint $\mathbf{1}^T \boldsymbol{\lambda} = 1$ to the dual problem (CCR$_D$) and thus, the variable u_o appears in the primal problem given by the problem (BCC):

$$\underset{\boldsymbol{\mu}, \boldsymbol{\nu}, u_o}{\text{maximize}} \qquad \sum_{k=1}^{p} \mu_k y_{ko} - u_o \qquad\qquad\qquad \text{(BCC)}$$

$$\text{subject to} \qquad \sum_{i=1}^{m} \nu_i x_{io} = 1,$$

$$\sum_{k=1}^{p} \mu_k y_{kj} - \sum_{i=1}^{m} \nu_i x_{ij} - u_o \leqq 0, \quad j = 1, \ldots, q,$$

$$\mu_k \geqq \varepsilon, \quad k = 1, \ldots, p,$$

$$\nu_i \geqq \varepsilon, \quad i = 1, \ldots, m.$$

The dual problem (BCC$_D$) to the problem (BCC) is formulated as follows:

$$\underset{\theta, \boldsymbol{\lambda}, \boldsymbol{s}_x, \boldsymbol{s}_y}{\text{minimize}} \qquad \theta - \varepsilon(\mathbf{1}^T \boldsymbol{s}_x + \mathbf{1}^T \boldsymbol{s}_y) \qquad\qquad \text{(BCC}_D)$$

$$\text{subject to} \qquad \boldsymbol{X}\boldsymbol{\lambda} - \theta \boldsymbol{x}_o + \boldsymbol{s}_x = \mathbf{0},$$

$$\boldsymbol{Y}\boldsymbol{\lambda} - \boldsymbol{y}_o - \boldsymbol{s}_y = \mathbf{0},$$

$$\mathbf{1}^T \boldsymbol{\lambda} = 1,$$

$$\boldsymbol{\lambda} \geqq \mathbf{0}, \quad \boldsymbol{s}_x \geqq \mathbf{0}, \quad \boldsymbol{s}_y \geqq \mathbf{0}.$$

The presence of the constraint $\mathbf{1}^T \boldsymbol{\lambda} = 1$ in the dual problem (BCC$_D$) yields that the production possibility set in BCC model is the *convex hull* generated by the observed data (see Fig. 3.12).

Adding the constraints $\lambda_j \in \{0, 1\}$, $j = 1, \ldots, q$, to the problem (BCC$_D$), FDH model in which the production possibility set is a *free disposable hull* (FDH) is given by Tulkens [151] is formulated as follows:

$$\underset{\theta, \boldsymbol{\lambda}, \boldsymbol{s}_x, \boldsymbol{s}_y}{\text{minimize}} \qquad \theta - \varepsilon(\mathbf{1}^T \boldsymbol{s}_x + \mathbf{1}^T \boldsymbol{s}_y) \qquad\qquad \text{(FDH}_D)$$

$$\text{subject to} \qquad \boldsymbol{X}\boldsymbol{\lambda} - \theta \boldsymbol{x}_o + \boldsymbol{s}_x = \mathbf{0},$$

$$\boldsymbol{Y}\boldsymbol{\lambda} - \boldsymbol{y}_o - \boldsymbol{s}_y = \mathbf{0},$$

$$\mathbf{1}^T \boldsymbol{\lambda} = 1; \quad \lambda_j \in \{0, 1\} \ j = 1, \ldots, q,$$

$$\boldsymbol{\lambda} \geqq \mathbf{0}, \quad \boldsymbol{s}_x \geqq \mathbf{0}, \quad \boldsymbol{s}_y \geqq \mathbf{0}.$$

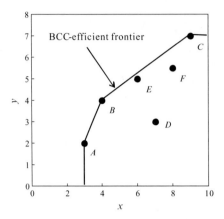

Fig. 3.12 Efficient frontier by BCC model

However, here, it is seen that the problem (FDH$_D$) is a mixed integer programming problem, and hence the traditional linear optimization methods cannot apply to it. An optimal solution is obtained by means of a simple vector comparison procedure to the end. For DMUo, the optimal solution θ^* to the problem (FDH$_D$) is equal to the value R_o^* defined by

$$R_o^* = \min_{j \in D(o)} \max_{i=1,\cdots m} \left\{ \frac{x_{ij}}{x_{io}} \right\}, \tag{3.3}$$

where $D(o) = \{\, j \mid \boldsymbol{x}_j \leq \boldsymbol{x}_o \text{ and } \boldsymbol{y}_j \geq \boldsymbol{y}_o, \ j = 1, \ldots, q \,\}$. R_o^* is substituted for θ^* as the efficiency of DMUo in FDH model (see Fig. 3.13).

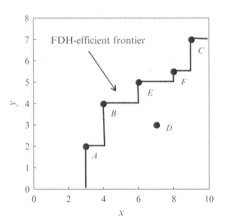

Fig. 3.13 Efficient frontier by FDH model

3.2.4 DEA Method

In order to evaluate quantitatively the fitness of each individual, Arakawa et al. [4] suggested a method using data envelopment analysis (we call this method *DEA method* in this book).

In DEA method, the fitness of an individual $\bm{x}^o$, $o = 1,\ldots,q$ in GA is evaluated by the optimal value to the following problem:

$$\operatorname*{minimize}_{\nu} \quad \frac{1}{\theta} = \sum_{i=1}^{m} \nu_i f_i(\bm{x}^o) \qquad (\mathrm{DEA}_{fit})$$

$$\text{subject to} \quad \sum_{i=1}^{m} \nu_i f_i(\bm{x}^j) \geqq 1, \ j = 1,\ldots,q,$$

$$\nu_i \geqq \varepsilon, \ i = 1,\ldots,m.$$

The optimal value θ to the problem (DEA_{fit}) represents the *relative degree* how close $\bm{f}(\bm{x}^o)$ is to DEA-efficient frontier. As shown in Fig. 3.14, the fitness of an individual F is the ratio of OP to OF, and the fitness of an individual G is the ratio of OQ to OG. Thus, by using DEA method, the individuals such as E and G far from the DEA-efficient frontier have low fitness, and hence can be removed at a relatively early generation. However, since DEA-efficient frontier is always convex,[5] DEA method may not generate the individuals in the sunken (or non-convex) part of Pareto frontier.

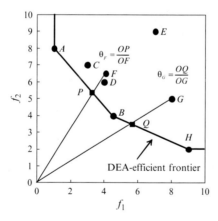

Fig. 3.14 DEA method

---

[5] Let E be the Pareto frontier set in $\mathbb{R}^m$ and let $\mathbb{R}^m_+$ be the nonnegative orthant in the objective space. Then we say the *Pareto frontier to be convex* if $(E + \mathbb{R}^m_+)$ is a convex set. Otherwise, the Pareto frontier is said to be nonconvex.

3.3 Fitness Evaluation Using GDEA

It should be noted that while DEA-efficient frontiers by CCR model [18] or BCC model [6] are convex, DEA-efficient frontier by FDH model [151] is nonconvex, since the free disposable hull of the given data set becomes nonconvex. Yun–Nakayama–Tanino [162] suggested a *generalized data envelopment analysis* (GDEA) which includes the existing DEA models such as CCR, BCC and FDH models by varying the value of parameter.

3.3.1 GDEA Model

First, we describe GDEA model. Assume that there exist q decision making units (DMUs) consuming m inputs and producing p outputs. Denote by $y_{kj} > 0$ and $x_{ij} > 0$ the amount of output k and that of input i generated by DMUj, respectively. Then, GDEA efficiency of DMUo, $o = 1, \ldots, q$ is judged by solving the following problem:

$$\underset{\Delta, \mu, \nu}{\text{maximize}} \quad \Delta \qquad\qquad\qquad\qquad (\text{GDEA})$$

$$\text{subject to} \quad \Delta \leqq \tilde{d}_j + \alpha \left(\sum_{k=1}^{p} \mu_k(y_{ko} - y_{kj}) + \sum_{i=1}^{m} \nu_i(-x_{io} + x_{ij}) \right),$$

$$j = 1, 2, \ldots, q,$$

$$\sum_{k=1}^{p} \mu_k + \sum_{i=1}^{m} \nu_i = 1,$$

$$\mu_k \geqq \varepsilon, \quad k = 1, \ldots, p,$$

$$\nu_i \geqq \varepsilon, \quad i = 1, \ldots, m,$$

where $\tilde{d}_j := \underset{\substack{k=1,\ldots,p \\ i=1,\ldots,m}}{\max} \{\nu_k (y_{ko} - y_{kj}), \mu_i (-x_{io} + x_{ij})\}$,[6] α is a given parameter.

GDEA efficiency[7] with specified values of the parameter α has some relation with conventional DEA efficiency. For a given α, DMUo is defined to be *GDEA-efficient* if the optimal value of the problem (GDEA) is equal to zero. The following properties between DEA efficiencies and GDEA efficiency are established [162].

---

[6] The formula means that $\tilde{d}_j$ is the value of multiplying the maximal component of $(y_{1o} - y_{1j}, \cdots, y_{po} - y_{pj}, -x_{1o} + x_{1j}, \cdots, -x_{mo} + x_{mj})$ by its corresponding weight. For example, if $(y_{1o} - y_{1j}, -x_{1o} + x_{1j}) = (2, -1)$, then $\tilde{d}_j = 2\mu_1$.

[7] The efficiency based on GDEA model is called *GDEA efficiency* or α *efficiency*.

3.3 Fitness Evaluation Using GDEA

Theorem 3.1. *DMUo is FDH-efficient if and only if DMUo is GDEA-efficient for sufficiently small $\alpha > 0$.*

Theorem 3.2. *DMUo is BCC-efficient if and only if DMUo is GDEA-efficient for sufficiently large $\alpha > 0$.*

Theorem 3.3. *Add the constraint $\sum_{k=1}^{p} \mu_k y_{ko} = \sum_{i=1}^{m} \nu_i x_{io}$ to the problem (GDEA). Then DMUo is CCR-efficient if and only if DMUo is GDEA-efficient for sufficiently large $\alpha > 0$.*

As is stated in the above theorems, various kinds of DEA-efficient frontiers are obtained by changing the value of parameter α in the model (GDEA).

Figure 3.15 shows various GDEA-efficient frontiers for several values of α. DMUs on the GDEA-efficient frontier are GDEA-efficient (or α-efficient). It

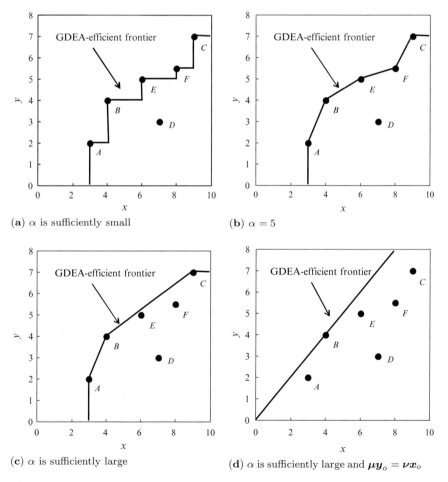

(a) α is sufficiently small

(b) $\alpha = 5$

(c) α is sufficiently large

(d) α is sufficiently large and $\boldsymbol{\mu y}_o = \boldsymbol{\nu x}_o$

Fig. 3.15 GDEA-efficient frontiers by the parameter α in GDEA model

is seen also from Fig. 3.15 that as the value of α becomes sufficiently large, GDEA-efficient frontier changes from the form of steps into the straight line. That is, GDEA model can generate the conventional DEA-efficient frontiers such as CCR-, BCC- and FDH-efficient frontiers depending on the value of α.

3.3.2 GDEA Method

In order to overcome the shortcomings of ranking methods and DEA method, Yun et al. [160] suggested to evaluate the fitness of the individual $\boldsymbol{x}^o$ in GA using GDEA model as follows:

$$\underset{\Delta,\,\boldsymbol{\nu}}{\text{maximize}} \quad \Delta \qquad\qquad\qquad\qquad (\text{GDEA}_{fit})$$

$$\text{subject to} \quad \Delta \leqq \tilde{d}_j - \alpha \sum_{i=1}^{m} \nu_i(f_i(\boldsymbol{x}^o) - f_i(\boldsymbol{x}^j)), \;\; j = 1, \ldots, q,$$

$$\sum_{i=1}^{m} \nu_i = 1,$$

$$\nu_i \geqq \varepsilon, \;\; i = 1, \ldots, m,$$

where $\tilde{d}_j = \underset{i=1,\ldots,m}{\max}\left\{ \nu_i\left(-f_i(\boldsymbol{x}^o) + f_i(\boldsymbol{x}^j)\right)\right\}$ and α is a value to decrease monotonically as the iteration proceeds.[8]

The degree of GDEA efficiency for an individual $\boldsymbol{x}^o$ is given by the optimal value Δ^* to the problem (GDEA_{fit}), and is used in the fitness evaluation. Therefore, the selection of an individual to survive in the subsequent generations is determined by the degree of GDEA efficiency, i.e., if Δ^* equals to zero, the individual remains at the next generation. With making the best use of the stated properties of GDEA, it is possible to keep merits of ranking methods and the method using DEA, and at the same time, to overcome the shortcomings of existing methods. Namely, taking a large α can remove individuals which are located far from GDEA-efficient frontier, and taking a small α can generate a nonconvex part of Pareto frontier (see Fig. 3.16).

---

[8] For example, α can be given by

$$\alpha(t) := \omega \cdot \exp(-\beta \cdot t), \;\; t = 0, 1, \ldots, N,$$

where ω, β and N are positive fixed numbers. $\omega\ (= \alpha(0))$ is determined to be sufficiently large as $10, 10^2$ and 10^3. N is the generation number. For given ω and N, β is decided by solving the equation $\alpha(N) = \omega \cdot \exp(-\beta \cdot N) \fallingdotseq 0$.

3.4 Comparisons of Several Fitness Evaluations

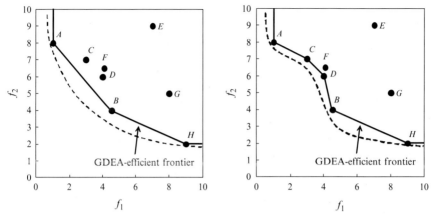

(a) Initial generation: α is sufficiently large (b) Intermediate generation: $\alpha = 10$

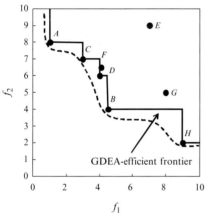

(c) Final generation: α is sufficiently small

Fig. 3.16 Geometric interpretation of α in (GDEA$_{fit}$): *points* are individuals, the *solid line* is GDEA-efficient frontier, and the *dotted line* is an approximate Pareto frontier

3.4 Comparisons of Several Fitness Evaluations

For comparing the methods for fitness evaluation which was introduced in the previous sections, consider three examples with two objective functions as below.

Example 3.2.

$$\begin{aligned}
\minimize_{x} \quad & \boldsymbol{f}(\boldsymbol{x}) := (x_1,\ x_2)^T \\
\text{subject to} \quad & (x_1 - 2)^2 + (x_2 - 2)^2 - 4 \leqq 0, \\
& x_1 \geqq 0,\ x_2 \geqq 0.
\end{aligned}$$

Example 3.3.

$$\underset{x}{\text{minimize}} \qquad f(x) := (2x_1 - x_2, \ -x_1)^T$$

$$\text{subject to} \qquad (x_1 - 1)^3 + x_2 \leqq 0,$$

$$x_1 \geqq 0, \ x_2 \geqq 0.$$

Example 3.4.

$$\underset{x}{\text{minimize}} \qquad f(x) := (x_1, \ x_2)^T$$

$$\text{subject to} \qquad x_1^3 - 3x_1 - x_2 \leqq 0,$$

$$x_1 \geqq -1, \ x_2 \leqq 2.$$

The Pareto frontier is convex for Example 3.2, and nonconvex for both Examples 3.3 and 3.4. The parameters in GA and the problem (GDEA$_{fit}$) are as follows:

1. Generation number: 30
2. Population size: 100
3. Crossover rate: 1
4. Crossover point: 1 or 2 point
5. Mutation rate: 0.05
6. $\alpha(t) = 50 \times \exp(-0.2 \times t), \ t = 0, \ldots, 30$

The elitist preserving selection [50] is adopted. Figures 3.17–3.19 show the results by (a) ranking method by Fonseca–Flemming, (b) DEA method, and (c) GDEA method. Here, symbol ○ represents all nondominated individuals generated at intermediate generations. Note here that nondominated individual depends on the domination structure of each method, for example, individual with the rank 1 in ranking method, DEA-efficient one in DEA method, and GDEA-efficient one in GDEA method.

As seen from the results, although ranking method generates relatively many nondominated individuals than the other methods, many individuals among them do not become finally nondominated solution at the final generation. While, the results by DEA method show an opposite tendency to ranking method. Moreover, DEA method is not suitable for cases with nonconvex Pareto frontiers, because of not generating nondominated solutions in the sunken part of Pareto frontier. GDEA method has the results which seems to overcome the weakness of ranking method and DEA method, and provide good nondominated solutions in terms of both the convergence and the diversity of population.

Particularly, it should be noted in ranking method that nondominated individuals obtained at intermediate generations are often finally not nondominated solutions. In many practical problems, we do not know when to

3.4 Comparisons of Several Fitness Evaluations

stop GA implementation in advance, and usually, the implementation is terminated at a relatively early generation due to the time limitation. It is an important requirement, therefore, that nondominated individuals at intermediate generations should be final solutions. From this point of view, it can be concluded that the methods using GDEA and DEA have a desirable performance.

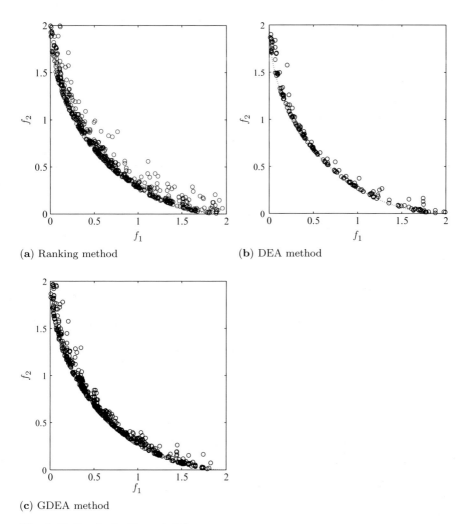

Fig. 3.17 Results for Example 3.2

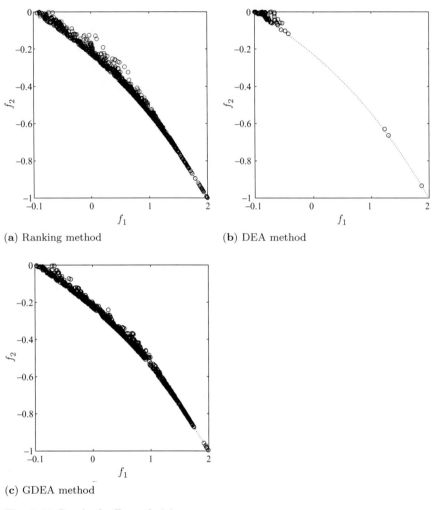

Fig. 3.18 Results for Example 3.3

3.4 Comparisons of Several Fitness Evaluations

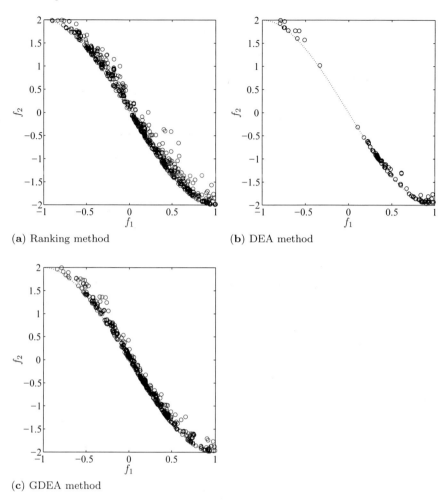

(a) Ranking method

(b) DEA method

(c) GDEA method

Fig. 3.19 Results for Example 3.4

Chapter 4
Multiobjective Optimization and Computational Intelligence

Computational intelligence is a comprehensive name embracing not only artificial intelligence but also several recent topics such as artificial neural networks, evolutionary methods, metaheuristics using swarm intelligence (e.g., particle swarm optimization, ant colony optimization), fuzzy logic and so on. Computational intelligence techniques are now widely applied in various practical fields. In this book also, we use several techniques of computational intelligence for sequential multiobjective optimization. This chapter will be concerned with machine learning which is utilized commonly for knowledge acquisition. Typical approaches to machine learning are (1) to find an explicit rule as *if–then rule* and (2) to judge newly observed data by an implicit rule which is usually represented as a nonlinear function. Well-known ID3 (recently C5.0) [119] and CART [14] belong to the former category. On the other hand, artificial neural networks and mathematical programming approaches belong to the latter category. In this book, we focus on the latter category.

4.1 Machine Learning

Typically machine learning is explained along pattern classification problems and regression problems.

Firstly, *pattern classification* problems are described as follows: Let X be a space of conditional attributes. For binary classification problems, the value of $+1$ or -1 is assigned to each pattern $\boldsymbol{x}_i \in X$ according to its class $\mathcal{A}$ or $\mathcal{B}$. The aim of machine learning is to predict which class newly observed patterns belong to on the basis of the given training data set $(\boldsymbol{x}_i, y_i)$, $i=1,\ldots,\ell$, where $y_i = +1$ or -1. This is performed by finding a *discriminant function* $f(\boldsymbol{x})$ such that $f(\boldsymbol{x}) \geqq 0$ for $\boldsymbol{x} \in \mathcal{A}$ and $f(\boldsymbol{x}) < 0$ for $\boldsymbol{x} \in \mathcal{B}$.

H. Nakayama et al., *Sequential Approximate Multiobjective Optimization*
Using Computational Intelligence, Vector Optimization,
DOI 10.1007/978-3-540-88910-6_4, © 2009 Springer-Verlag Berlin Heidelberg

Linear discriminant functions, in particular, can be expressed by the following linear form

$$f(\boldsymbol{x}) = \boldsymbol{w}^T \boldsymbol{x} + b$$

with the property

$$\boldsymbol{w}^T \boldsymbol{x} + b \geq 0 \qquad \text{for} \qquad \boldsymbol{x} \in \mathcal{A},$$
$$\boldsymbol{w}^T \boldsymbol{x} + b < 0 \qquad \text{for} \qquad \boldsymbol{x} \in \mathcal{B}.$$

Secondly, regression problems are aimed at predicting the value of some functions f at the newly observed point $\boldsymbol{x}_0$, namely $y_0 = f(\boldsymbol{x}_0)$, on the basis of some existing sample data $(\boldsymbol{x}_1, y_i), \ldots, (\boldsymbol{x}_\ell, y_\ell)$. Regression analysis has been commonly developed in traditional statistics.

In either cases, our aim is finally to find an approximate function which can provide as a correct judgment as possible for newly observed points on the basis of existing observations.

For this purpose, artificial neural networks have been widely applied. However, the back propagation method is reduced to nonlinear optimization with multiple local optima, and hence difficult to apply to large-scale problems. Another drawback in the back propagation method is in the fact that it is difficult to change the structure adaptively according to the change of environment in *incremental learning*.

One of promising techniques for machine learning is *radial basis function networks* (RBFN). The learning of RBFN is usually reduced to solving linear equations. Therefore, its learning is very fast in general. Moreover, it is easy to make incremental learning in RBFN, because the structure of networks is changeable according to situations relatively easily. RBFN has been applied to a variety of practical fields (see, e.g., [54]).

Recently, *support vector machine* (SVM) is attracting interest of researchers, in particular, people who are engaged in mathematical programming, because it is reduced to quadratic programming (QP) or linear programming (LP). One of main features in SVM is that it is a linear classifier with maximal margin on the feature space through nonlinear mappings defined implicitly by kernels in Hilbert space. The idea of *maximal margin* in linear classifiers is intuitive, and its reasoning in connection with perceptrons was given in early 1960s, e.g., Novikoff [110]. The maximal margin is effectively applied for discrimination analysis using mathematical programming, e.g., multisurface method (MSM) by Mangasarian [83]. Later, linear classifiers with maximal margin were formulated as linear goal programming, and extensively studied through 1980s to the beginning of 1990s. The pioneering work was given by Freed–Glover [45], and a good survey can be seen in Erenguc–Koehler [38].

This chapter introduces RBFN first, and then SVMs and several extensions using techniques of multiobjective programming (MOP) and goal programming (GP).

4.1 Machine Learning 75

4.1.1 Learning and Generalization

In this section, we shall give a rough sketch of the outline of fundamental concepts in machine learning along regression problems. For more details, e.g., see the references [10, 22, 53, 55, 133].

Let y be an observation of the output of f at x with random error ε whose mean is 0. Then we have

$$y = f(x) + \varepsilon.$$

Let us construct an approximate function (called a *metamodel* in later sections) $\hat{f}$ by deciding the "best" parameter w for parameterized $\hat{f}(x, w)$ on the basis of observed data (x_i, y_i), $i = 1, \ldots \ell$, which are drawn from the distribution of the joint probability density function

$$p(x, y) = p(x)p(y|x).$$

The estimation error between the observation y and the output of the metamodel $\hat{f}(x, w)$ at a point x is given by a so-called *loss function* $L(y, \hat{f}(x, w))$. For example, the most popular loss function is given by

$$L(y, \hat{f}(x, w)) = (y - \hat{f}(x, w))^2.$$

The expected value of the loss is referred to as the *risk* which is given by

$$R(w) = \int L(y, \hat{f}(x, w))p(x, y)dxdy.$$

Using the loss function of squared error, we have the following relation

$$
\begin{aligned}
R(w) &= \int (y - \hat{f}(x, w))^2 p(x, y)dxdy \\
&= \int (y - f(x) + f(x) - \hat{f}(x, w))^2 p(x, y)dxdy \\
&= \int (y - f(x))^2 p(x, y)dxdy + \int (\hat{f}(x, w) - f(x))^2 p(x, y)dxdy \\
&\quad + 2 \int (y - f(x))(f(x) - \hat{f}(x, w))p(x, y)dxdy \\
&= \int (y - f(x))^2 p(x, y)dxdy + \int (\hat{f}(x, w) - f(x))^2 p(x, y)dxdy \\
&= \int E_\varepsilon(\varepsilon^2|x)p(x)dx + \int (\hat{f}(x, w) - f(x))^2 p(x, y)dxdy,
\end{aligned}
$$

where $E_\varepsilon(\varepsilon^2|x) = \int \varepsilon^2 p(y|x)dy$.

Since the first term is not affected by the parameter w, minimizing the prediction risk is equivalent to finding the parameter w which yields the minimal discrepancy between the metamodel $\hat{f}(x, w)$ and the model $f(x)$.

This effort to find a parameter w giving the minimal prediction risk (i.e., as in traditional regression) or equivalently giving the best metamodel is called *learning* in this book.

Since the density function is not known, there are several approaches to estimating an optimal w^* minimizing the risk functional on the basis of observed data (x_i, y_i), $i = 1, \ldots, \ell$. One of them is to estimate the density which leads to the *maximum likelihood*. Another popular approach is to use the *empirical risk* R_{emp} which takes the expectation of $R(w)$ over the given data set, namely

$$R_{\text{emp}} = \frac{1}{\ell} \sum_{i=1}^{\ell} L(y_i, \hat{f}(x_i, w)).$$

It is known that the optimal solution $\hat{w}$ minimizing R_{emp} converges to the solution w^* minimizing $R(w)$ as $\ell \to \infty$ in probability by *induction rule* [54, 153].

On the other hand, note that our final goal is to predict the value of $f(x_0)$ at an untried point x_0 as correctly as possible. This is called *generalization*. The *expected prediction error* at an untried point x_0 using squared loss is given by

$$E\left(\left(\hat{f}(x_0, w) - f(x_0) \right)^2 \right) = E\left(\left(\hat{f}(x_0, w) - E(\hat{f}(x_0, w)) \right)^2 \right)$$
$$+ \left(E(\hat{f}(x_0, w)) - f(x_0) \right)^2.$$

The first term in the right-hand side of the above formula is the variance of $\hat{f}(x_0, w)$, and the second one is the squared bias which measures the difference between the estimate and its true value. Taking the global average over x, we define the *mean squared error*, *bias* and *variance* as follows:

$$\text{MSE}(\hat{f}(x, w)) = \int E\left(\left(\hat{f}(x, w) - f(x) \right)^2 \right) p(x) dx,$$
$$\text{bias}^2(\hat{f}(x, w)) = \int \left(E(\hat{f}(x, w)) - f(x) \right)^2 p(x) dx,$$
$$\text{Var}(\hat{f}(x, w)) = \int E\left(\left(\hat{f}(x, w) - E(\hat{f}(x, w)) \right)^2 \right) p(x) dx.$$

Summarizing, the above relation can be rewritten as

$$\text{MSE}(\hat{f}(x, w)) = \text{bias}^2(\hat{f}(x, w)) + \text{Var}(\hat{f}(x, w)).$$

In general, decreasing the bias overly by making the model $\hat{f}$ too much complex yields a large amount of variance. This is called the *bias–variance trade-off*. This should be taken into account when selecting our model.

4.1.2 The Least Square Method

Now consider a traditional regression problem: find the best estimation of parameter w of metamodel $\hat{f}(x, w)$ on the basis of observed data (x_i, y_i) with random error such that

$$y_i = \hat{f}(x_i, w) + \varepsilon_i,$$

where the error ε_i is independent of x and drawn by the distribution of a known density $p_\varepsilon(\varepsilon)$.

Letting $Z = \{(x_i, y_i), \ i = 1, \ldots, \ell\}$, and assuming the error is normally distributed with zero mean and a fixed variance σ, the *likelihood* is given by

$$P(Z|w) = \sum_{i=1}^{\ell} \ln p_\varepsilon(y_i - \hat{f}(x_i, w))$$

$$= -\frac{1}{2\sigma^2} \sum_{i=1}^{\ell} (y_i - \hat{f}(x_i, w))^2 - \ell \ln(\sqrt{2\pi}\sigma).$$

Therefore, maximizing the likelihood above is equivalent to minimizing the following the *empirical risk* using squared error

$$R_{\text{emp}} = \frac{1}{\ell} \sum_{i=1}^{\ell} (y_i - \hat{f}(x_i, w))^2.$$

Note here that the assumption of Gaussian noise is crucial.

From now on, let us concentrate our consideration on metamodels linear in parameter w using the basis functions $h_j, \ j = 1, \ldots, m$,

$$\hat{f}(x, w) = \sum_{j=1}^{m} w_j h_j(x).$$

Suppose that for observed data $(x_i, y_i), \ i = 1, \ldots, \ell$, we have

$$y_i = \sum_{j=1}^{m} w_j h_j(x_i) + \varepsilon_i, \ i = 1, \ldots, \ell, \tag{4.1}$$

where ε_i are random error terms with the mean $E(\varepsilon_i) = 0$, the same variance $\text{var}(\varepsilon_i) = \sigma^2$ and also uncorrelated so that their covariance is zero.

Then, the squared error is denoted by

$$\mathcal{E}(w) = \sum_{i=1}^{\ell} \left(y_i - \sum_{j=1}^{m} w_j h_j(x_i) \right)^2.$$

At this event, if the observation y_i at x_i is multiple, we can take its average as y_i.

Letting $h = (h_1, \ldots, h_m)^T$, $H = (h(x_1), \ldots, h(x_\ell))^T$, $y = (y_1, \ldots, y_\ell)^T$ and $\varepsilon = (\varepsilon_1, \ldots, \varepsilon_\ell)^T$, the regression model (4.1) can be rewritten by

$$y = Hw + \varepsilon, \quad \varepsilon \sim N(\mathbf{0}, \sigma^2 I_\ell),$$

where I_ℓ denotes the ℓ-order identity matrix. With this matrix notations, the squared error can be rewritten by

$$\mathcal{E}(w) = ||y - Hw||^2 = (y - Hw)^T (y - Hw). \tag{4.2}$$

Differentiating the above $\mathcal{E}(w)$ with respect to w, the least square estimator $\hat{w}$ satisfies

$$H^T H \hat{w} = H^T y.$$

This equation is called the *normal equation*. Therefore, assuming $H^T H$ to be nonsingular, the *least square estimator* $\hat{w}$ is given by

$$\hat{w} = (H^T H)^{-1} H^T y. \tag{4.3}$$

Remark 4.1. In order to avoid the invalidity of (4.3) due to the singularity of the matrix $H^T H$, we penalize the error function (4.2) as follows:

$$\mathcal{E}_1(w) = ||y - Hw||^2 + \lambda ||w||^2.$$

For the above $\mathcal{E}_1$, the normal equation is given by

$$(H^T H + \lambda I_m) \hat{w} = H^T y,$$

where I_m is the m-order identity matrix. Now we have the solution

$$\hat{w} = (H^T H + \lambda I_m)^{-1} H^T y,$$

This technique is called *regularization* for ill-posed problems. The regression using the above regularization is called *ridge regression*.

Example 4.1. For $x \in \mathbb{R}^n$, consider a linear regression model appearing in standard statistics:

$$y_i = \beta_0 + \beta_1 x_{i1} + \cdots + \beta_n x_{in} + \varepsilon_i, \quad i = 1, \ldots, \ell,$$

where $\beta_0, \beta_1, \ldots, \beta_n$ are unknown parameters. In this case, we interpret in our general model

$$h_1(x) = 1, \ h_2(x) = x_1, \ \ldots, \ h_{n+1}(x) = x_n.$$

4.2 Radial Basis Function Networks

Then we have

$$H = \begin{pmatrix} 1 & x_{11} & x_{12} & \cdots & x_{1n} \\ 1 & x_{21} & x_{22} & \cdots & x_{2n} \\ \vdots & \vdots & \vdots & \ddots & \vdots \\ 1 & x_{\ell 1} & x_{\ell 2} & \cdots & x_{\ell n} \end{pmatrix}.$$

Remark 4.2. Usually in such cases with polynomial basis functions in traditional statistics, the matrix H is mostly denoted as X.

4.2 Radial Basis Function Networks

Radial basis functions were originally often discussed from a viewpoint of *approximation* or *interpolation* [15, 116]. Consider the following function $\hat{f}(\boldsymbol{x})$ expressed as a linear combination of basis functions $h_j(\boldsymbol{x})$, $j = 1, \ldots, m$, which are regarded as outputs of hidden units in a framework of neural networks:

$$\hat{f}(\boldsymbol{x}) = \sum_{j=1}^{m} w_j h_j(\boldsymbol{x}), \tag{4.4}$$

where $\boldsymbol{x} \in \mathbb{R}^n$ is an input vector.

Radial basis function networks (RBFN) are neural networks whose output is given by $\hat{f}(\boldsymbol{x})$ with *radial basis functions* (RBF) $h_j(\boldsymbol{x})$, $j = 1, \ldots, m$ [55, 114]. Figure 4.1 shows the most basic construction of RBFN in a single hidden-layer network.

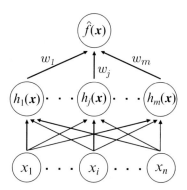

Fig. 4.1 Radial basis function networks

Radial functions have the feature that their outputs are monotonically increasing/decreasing with respect to the distance of an input vector $\boldsymbol{x}$ from the so-called *center*; a radial function h_j with the center $c_j \in \mathbb{R}^n$ scaled by a metric $\mathbf{M}_j$ is given by

$$h_j(\boldsymbol{x}) = \phi\left((\boldsymbol{x} - \boldsymbol{c}_j)^T \mathbf{M}_j^{-1}(\boldsymbol{x} - \boldsymbol{c}_j)\right),$$

where ϕ is some monotonic function.

There are several types of radial function such as Gauss, multiquadric, inverse multiquadric, pseudocubic, Cauchy and thin-plate-spline functions. One of most commonly used radial functions is the Gauss function defined by

$$h_j(\boldsymbol{x}) = \exp\left(-\frac{\|\boldsymbol{x} - \boldsymbol{c}_j\|^2}{2r_j^2}\right),$$

where $\boldsymbol{c}_j$ and r_j is, respectively, called the *center* and the *width* (or *radius*) of the jth Gauss function (Fig. 4.2).

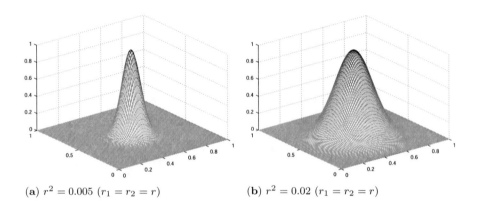

(a) $r^2 = 0.005$ ($r_1 = r_2 = r$) (b) $r^2 = 0.02$ ($r_1 = r_2 = r$)

Fig. 4.2 Gauss functions

Suppose that the linear model is given by (4.4), and the training data $(\boldsymbol{x}_i, y_i)$, $i = 1, \ldots, \ell$. Then, the learning of RBFN is usually performed by minimizing the following *cost function* $C(\boldsymbol{w})$ in which the penalty for *weight decay* (the second term) is added to the sum of squared errors (the first term),

$$C(\boldsymbol{w}) = \sum_{i=1}^{\ell} \left(y_i - \hat{f}(\boldsymbol{x}_i)\right)^2 + \sum_{j=1}^{m} \lambda_j w_j^2 \longrightarrow \min, \qquad (4.5)$$

where λ_j, $j = 1, \ldots, m$ are regularization parameters.

That is, the learning of RBFN is reduced to finding the optimal weight vector $\boldsymbol{w}$ in the cost function (4.5), represented finally by the formula

$$\boldsymbol{w} = \left(H^T H + \Lambda\right)^{-1} H^T \boldsymbol{y}, \qquad (4.6)$$

4.2 Radial Basis Function Networks

where H, Λ and $\boldsymbol{y}$ are the following:

$$H = \begin{bmatrix} h_1(\boldsymbol{x}_1) & h_2(\boldsymbol{x}_1) & \cdots & h_m(\boldsymbol{x}_1) \\ h_1(\boldsymbol{x}_2) & h_2(\boldsymbol{x}_2) & \cdots & h_m(\boldsymbol{x}_2) \\ \vdots & \vdots & \ddots & \vdots \\ h_1(\boldsymbol{x}_\ell) & h_2(\boldsymbol{x}_\ell) & \cdots & h_m(\boldsymbol{x}_\ell) \end{bmatrix},$$

$$\Lambda = \begin{bmatrix} \lambda_1 & 0 & \cdots & 0 \\ 0 & \lambda_2 & \cdots & 0 \\ \vdots & \vdots & \ddots & \vdots \\ 0 & 0 & \cdots & \lambda_m \end{bmatrix},$$

$$\boldsymbol{y} = \begin{bmatrix} y_1 \\ y_2 \\ \vdots \\ y_\ell \end{bmatrix}.$$

Example 4.2. Using the Gauss basis function in RBFN, Fig. 4.3 shows the classification results for several values of parameters r and λ, respectively.

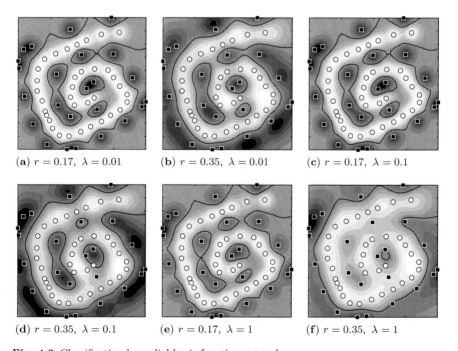

(a) $r = 0.17$, $\lambda = 0.01$ (b) $r = 0.35$, $\lambda = 0.01$ (c) $r = 0.17$, $\lambda = 0.1$

(d) $r = 0.35$, $\lambda = 0.1$ (e) $r = 0.17$, $\lambda = 1$ (f) $r = 0.35$, $\lambda = 1$

Fig. 4.3 Classification by radial basis function networks

Example 4.3. The results for regression by using RBFN with several values of r and λ are shown in Fig. 4.4.

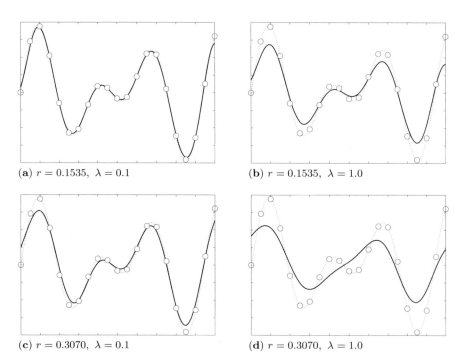

(a) $r = 0.1535$, $\lambda = 0.1$ (b) $r = 0.1535$, $\lambda = 1.0$

(c) $r = 0.3070$, $\lambda = 0.1$ (d) $r = 0.3070$, $\lambda = 1.0$

Fig. 4.4 Regression by radial basis function networks: the *solid line* depicts the predicted function, and the *dotted line* does the true function

As can be seen in Figs. 4.3 and 4.4, the width parameter r affects the smoothness of predicted (or discriminant) function. If the value of r is too large, the approximation (or discrimination) is not sufficient. On the other hand, if the value of r is too small, the approximate function may have poor generalization, although it can fit training data well. It has been observed that the value of r needs to be set small to some extent for highly nonlinear cases. The regularization parameter λ is a trade-off parameter between the weight decay (the second term) and the total error (the first term) in the cost function (4.5). Thus, the smaller the value of λ is, the more important the total error term becomes. It has been recognized that the value of λ in RBFN affects the robustness against noises.

Remark 4.3. It is important to decide moderately the parameters, such as the center c_j and the width (radius) r_j of Gauss basis function and λ in RBFN. The task for deciding appropriate values of parameters are regarded as one of model selection. Although there have been proposed several criteria such as AIC (*Akaike's information criterion*) [2], BIC (*Bayesian*

4.3 Support Vector Machines for Pattern Classification 83

information criterion) [10] for model selection, they are usually decided by *cross-validation* in practice as follows.

Divide the given data set D into a training data set D_T and a validation data set D_V, namely $D = D_T \cup D_V$. The basic idea of cross-validation is to decide the values of parameters by evaluating the generalization by changing the combination of the sets D_T and D_V variously. One of most popular method is to select D_V as a singleton, which is called *leave-one-out method*. For large scale data sets, however, the leave-one-out method is time consuming for checking the validation sufficiently well. Under this circumstance, split the original data set D into K roughly equal-sized subsets $D_1, \ldots, D_K$. Then take one of them, e.g., D_i, as the validation set D_V, while the rest $D_j, j = \{1, \ldots, K\}\backslash i$, are used as training data sets. By changing such i from 1 to K, we check the effectiveness of parameter values. This is called K-*fold cross-validation test*. Usually, $K = 5$ or 10 is recommended [53].

For RBFN using Gauss basis function, we often use the following simple estimate of the width (radius) r in common for all Gauss basis functions which is a slight modification of the one in [54]:

$$r = \frac{d_{max}}{\sqrt[n]{nm}}, \tag{4.7}$$

where d_{max} is the maximal distance among the input data, n is the dimension of the input data and m is the number of basis functions. Using (4.7), we have $r = 0.17$ in Fig. 4.3 and $r = 0.307$ in Fig. 4.4.

4.3 Support Vector Machines for Pattern Classification

Support vector machine (SVM) is gaining much popularity as one of effective methods for machine learning in recent years [25, 27, 133, 146]. In pattern classification problems with two class sets, it generalizes linear classifiers into high-dimensional feature spaces through nonlinear mappings defined implicitly by kernels in the Hilbert space so that it may produce nonlinear classifiers in the original input data space. Linear classifiers then are optimized to give the maximal margin separation between the classes. This task is performed by solving some type of mathematical programming such as quadratic programming (QP) or linear programming (LP).

On the other hand, from a viewpoint of mathematical programming for machine learning, the idea of maximal margin separation was employed in the multisurface method (MSM) suggested by Mangasarian in 1960s [83]. Also, linear classifiers using goal programming were developed extensively in 1980s.

In this section, including the conventional SVM models, we introduce a new family of SVM using multiobjective programming and goal programming (MOP/GP) techniques. Their effectiveness will be compared throughout several numerical experiments.

4.3.1 Hard Margin SVM

Support vector machine (SVM) was developed by Vapnik et al. [25, 153] (see also Cristianini and Shawe-Taylor [27], Schölkopf and Smola [133]) and its main features are:

1. SVM maps the original input data into a high-dimensional feature space by nonlinear mapping implicitly defined by kernels in the Hilbert space.
2. SVM finds linear classifiers with maximal margin on the feature space.
3. SVM provides an evaluation of the generalization ability using VC dimension.

Namely, in cases where training input data $x_1, \ldots, x_\ell \in X$ are not linearly separable, the original input space X is mapped to a *feature space* Z by some nonlinear transformation ϕ. Increasing the dimension of the feature space, it is expected that the mapped data set becomes linearly separable. Linear classifiers with *maximal margin* can be found in the feature space (see Fig. 4.5).

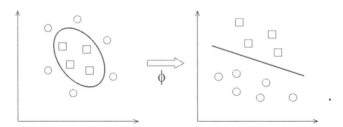

Fig. 4.5 Mapping of data from an original input space to a feature space

Letting $z_i = \phi(x_i)$ for the given training data (x_i, y_i), $i = 1, \ldots, \ell$, where $y_i = +1$ or -1, the separating hyperplane with maximal margin can be given by solving the following problem with the normalization $w^T z + b = \pm 1$ at points with the minimum *interior deviation* (see Fig. 4.6):

$$\begin{aligned}
&\underset{w,b}{\text{minimize}} & & \|w\| & & (\text{SVM}_{hard})_P \\
&\text{subject to} & & y_i\left(w^T z_i + b\right) \geqq 1, \ i = 1, \ldots, \ell.
\end{aligned}$$

4.3 Support Vector Machines for Pattern Classification

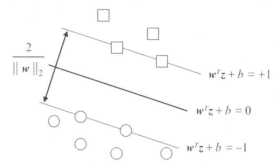

Fig. 4.6 Geometric interpretation of margin

Several kinds of norm are possible. When $\|w\|_2$ is used, the problem is reduced to quadratic programming, while the problem with $\|w\|_1$ or $\|w\|_\infty$ is reduced to linear programming (see, e.g., [84]).

Dual problem of $(\text{SVM}_{hard})_P$ with $\frac{1}{2}\|w\|_2^2$ is given by

$$\underset{\alpha}{\text{maximize}} \quad \sum_{i=1}^{\ell} \alpha_i - \frac{1}{2} \sum_{i,j=1}^{\ell} \alpha_i \alpha_j y_i y_j \phi(\bm{x}_i)^T \phi(\bm{x}_j) \quad (\text{SVM}_{hard})_D$$

$$\text{subject to} \quad \sum_{i=1}^{\ell} \alpha_i y_i = 0,$$

$$\alpha_i \geqq 0, \ i = 1, \ldots, \ell.$$

Using the *kernel function* $K(\bm{x}, \bm{x}')$ with the property

$$K(\bm{x}, \bm{x}') = \phi(\bm{x})^T \phi(\bm{x}'),$$

the problem $(\text{SVM}_{hard})_D$ can be reformulated as follows:

$$\underset{\alpha}{\text{maximize}} \quad \sum_{i=1}^{\ell} \alpha_i - \frac{1}{2} \sum_{i,j=1}^{\ell} \alpha_i \alpha_j y_i y_j K(\bm{x}_i, \bm{x}_j) \quad (\text{SVM}_{hard})$$

$$\text{subject to} \quad \sum_{i=1}^{\ell} \alpha_i y_i = 0,$$

$$\alpha_i \geqq 0, \ i = 1, \ldots, \ell.$$

Several kinds of kernel functions have been suggested: among them, inhomogeneous polynomial function

$$K(\bm{x}, \bm{x}') = (\bm{x}^T \bm{x}' + 1)^q$$

and Gaussian kernel function

$$K(\boldsymbol{x}, \boldsymbol{x}') = \exp\left(-\frac{||\boldsymbol{x} - \boldsymbol{x}'||^2}{2\sigma^2}\right)$$

are most popularly used.

Using the hard margin SVM with Gauss kernel function ($\sigma = 0.35$), Fig. 4.7 shows the result for Example 4.2. Here, the discriminant function $f(\boldsymbol{x}) = 0$ is depicted by solid line and support vectors by the symbol $*$.

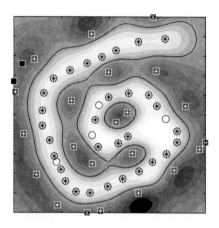

Fig. 4.7 Classification by hard margin SVM

4.3.2 MOP/GP Approaches to Pattern Classification

In 1981, Freed–Glover [45] suggested to get just a hyperplane separating two classes with as few misclassified data as possible by using goal programming (see also [38]).

Let ξ_i denote the *exterior deviation* which is a deviation from the hyperplane of a point $\boldsymbol{x}_i$ improperly classified. Similarly, let η_i denote the *interior deviation* which is a deviation from the hyperplane of a point $\boldsymbol{x}_i$ properly classified. Some of main objectives in this approach are as follows:

1. Minimize the maximum exterior deviation (decrease errors as much as possible).
2. Maximize the minimum interior deviation (i.e., maximize the margin).
3. Maximize the weighted sum of interior deviation.
4. Minimize the weighted sum of exterior deviation.

Although many models have been suggested, the one considering (3) and (4) in the above may be given by the following linear goal programming:

4.3 Support Vector Machines for Pattern Classification

$$\underset{\boldsymbol{w},b,\boldsymbol{\xi},\boldsymbol{\eta}}{\text{minimize}} \qquad \sum_{i=1}^{\ell} (h_i \xi_i - k_i \eta_i) \qquad\qquad (\text{MOP/GP}_1)$$

$$\text{subject to} \qquad y_i \left(\boldsymbol{x}_i^T \boldsymbol{w} + b\right) = \eta_i - \xi_i,$$

$$\xi_i, \ \eta_i \geqq 0, \ i = 1,\dots,\ell,$$

where since $y_i = +1$ or -1 according to $\boldsymbol{x}_i \in \mathcal{A}$ or $\boldsymbol{x}_i \in \mathcal{B}$, two equations $\boldsymbol{x}_i^T \boldsymbol{w} + b = \eta_i - \xi_i$ for $\boldsymbol{x}_i \in \mathcal{A}$ and $\boldsymbol{x}_i^T \boldsymbol{w} + b = -\eta_i + \xi_i$ for $\boldsymbol{x}_i \in \mathcal{B}$ can be reduced to the following one equation

$$y_i \left(\boldsymbol{x}_i^T \boldsymbol{w} + b\right) = \eta_i - \xi_i.$$

Here, h_i and k_i are positive constants. In order for ξ_i and η_i to have the meaning of the exterior deviation and the interior deviation, respectively, the condition $\xi_i \eta_i = 0$ for every $i = 1,\dots,\ell$ at the solution to (MOP/GP_1) must hold.

Lemma 4.1. *If $h_i > k_i$ for $i = 1,\dots,\ell$, then we have $\xi_i \eta_i = 0$ for every $i = 1,\dots,\ell$ at the solution to (MOP/GP$_1$).*

Proof. Easy in a similar fashion to Lemma 2.1. $\square$

It should be noted that the above formulation may yield some unacceptable solutions such as $\boldsymbol{w} = 0$ and unbounded solution. In the goal programming approach to linear classifiers, therefore, some appropriate normality condition must be imposed on $\boldsymbol{w}$ in order to provide a bounded nontrivial optimal solution. One of such normality conditions is $||\boldsymbol{w}|| = 1$.

If the classification problem is linearly separable, then using the normalization $||\boldsymbol{w}|| = 1$, the separating hyperplane $H : \boldsymbol{w}^T \boldsymbol{x} + b = 0$ with maximal margin can be given by solving the following problem [16]:

$$\underset{\boldsymbol{w},b,\eta}{\text{maximize}} \qquad \eta \qquad\qquad (\text{MOP/GP}_2)$$

$$\text{subject to} \qquad y_i \left(\boldsymbol{x}_i^T \boldsymbol{w} + b\right) \geqq \eta, \ i = 1,\dots,\ell,$$

$$||\boldsymbol{w}|| = 1.$$

However, this normality condition makes the problem to be of nonlinear optimization. Instead of maximizing the minimum interior deviation in (MOP/GP_2), we can use the following equivalent formulation with the normalization $\boldsymbol{x}^T \boldsymbol{w} + b = \pm 1$ at points with the minimum interior deviation [85]:

$$\underset{\boldsymbol{w},b}{\text{minimize}} \qquad ||\boldsymbol{w}|| \qquad\qquad (\text{MOP/GP}_2')$$

$$\text{subject to} \qquad y_i \left(\boldsymbol{x}_i^T \boldsymbol{w} + b\right) \geqq \eta, \ i = 1,\dots,\ell,$$

$$\eta = 1.$$

This formulation is the same as the problem $(\text{SVM}_{hard})_P$ of the basic SVM described in Sect. 4.3.1.

4.3.3 Soft Margin SVM

Separating two sets $\mathcal{A}$ and $\mathcal{B}$ completely is called the hard margin SVM, which tends to make overlearning. This implies the hard margin method is easily affected by noise. In order to overcome this difficulty, the *soft margin SVM* is introduced. The soft margin SVM allows some slight error which is represented by *slack variables* (exterior deviation) ξ_i, $i = 1, \ldots, \ell$ (Fig. 4.8).

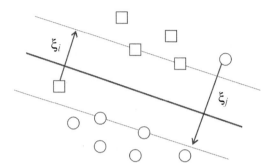

Fig. 4.8 Slack variable (exterior deviation)

Using the *trade-off parameter* C between minimizing $||\boldsymbol{w}||$ and minimizing $\sum_{i=1}^{\ell} \xi_i$, the formulation for the soft margin SVM is given by the following:

$$\begin{aligned}
\underset{\boldsymbol{w},b,\boldsymbol{\xi}}{\text{minimize}} \quad & \frac{1}{2}||\boldsymbol{w}||_2^2 + C\sum_{i=1}^{\ell} \xi_i & (\text{SVM}_{soft})_P\\
\text{subject to} \quad & y_i\left(\boldsymbol{w}^T \boldsymbol{z}_i + b\right) \geqq 1 - \xi_i, \ i = 1, \ldots, \ell,\\
& \xi_i \geqq 0, \ i = 1, \ldots, \ell.
\end{aligned}$$

Using a kernel function in the dual problem yields

$$\begin{aligned}
\underset{\boldsymbol{\alpha}}{\text{maximize}} \quad & \sum_{i=1}^{\ell} \alpha_i - \frac{1}{2}\sum_{i,j=1}^{\ell} \alpha_i \alpha_j y_i y_j K(\boldsymbol{x}_i, \boldsymbol{x}_j) & (\text{SVM}_{soft})\\
\text{subject to} \quad & \sum_{i=1}^{\ell} \alpha_i y_i = 0,\\
& 0 \leqq \alpha_i \leqq C, \ i = 1, \ldots, \ell.
\end{aligned}$$

For Example 4.2 which was also used in Sect. 4.3.1, the results by using the soft margin SVM are shown in Fig. 4.9.

4.3 Support Vector Machines for Pattern Classification 89

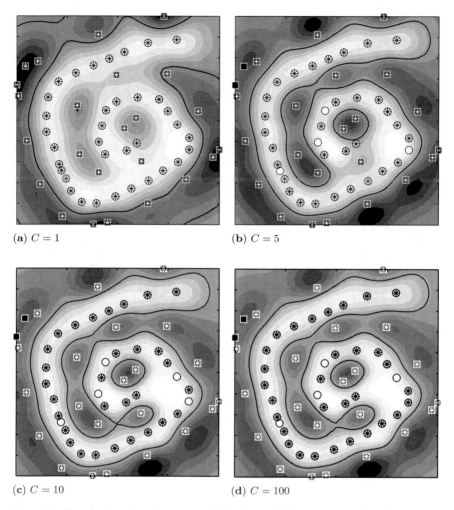

Fig. 4.9 Classification by soft margin SVM: the *solid line* represents the discriminant function and the *symbol* ∗ does support vectors

It is seen that the idea of soft margin SVM is the same as the goal programming approach to linear classifiers minimizing the sum of exterior deviation. This idea was used in an extension of MSM by Bennett [8]. Not only exterior deviations but also interior deviations can be considered in SVM. Such MOP/GP approaches to SVM are discussed by the authors and their coresearchers [5, 99, 101, 102, 108, 158, 163].

4.3.4 ν-SVM

Lately, ν-*support vector algorithm* was suggested by Schölkopf–Smola [132]:

$$\underset{\boldsymbol{w},b,\boldsymbol{\xi},\rho}{\text{minimize}} \qquad \frac{1}{2}||\boldsymbol{w}||_2^2 - \nu\rho + \frac{1}{\ell}\sum_{i=1}^{\ell}\xi_i \qquad (\nu\text{-SVM})_P$$

$$\text{subject to} \qquad y_i\left(\boldsymbol{w}^T\boldsymbol{z}_i + b\right) \geqq \rho - \xi_i,$$

$$\rho \geqq 0, \ \xi_i \geqq 0, \ i = 1,\ldots,\ell,$$

where $0 < \nu \leqq 1$ is a parameter.

Compared with the existing soft margin SVM, one of the differences is that the parameter C for slack variables does not appear, and another difference is that the new variable ρ appears in the above formulation. The problem (ν-SVM)$_P$ maximizes the variable ρ which corresponds to the minimum interior deviation, i.e., the minimum distance between the separating hyperplane and correctly classified points.

Note that this formulation is the same as the one taking into account the objectives (2) and (4) of goal programming stated in Sect. 4.3.2.

The Lagrangian dual problem to the problem (ν-SVM)$_P$ is as follows:

$$\underset{\boldsymbol{\alpha}}{\text{maximize}} \qquad -\frac{1}{2}\sum_{i,j=1}^{\ell} y_iy_j\alpha_i\alpha_j K\left(\boldsymbol{x}_i,\boldsymbol{x}_j\right) \qquad (\nu\text{-SVM})$$

$$\text{subject to} \qquad \sum_{i=1}^{\ell} y_i\alpha_i = 0,$$

$$\sum_{i=1}^{\ell} \alpha_i \geqq \nu,$$

$$0 \leqq \alpha_i \leqq \frac{1}{\ell}, \ i = 1,\ldots,\ell.$$

Figure 4.10 shows the results by ν-SVM for Example 4.2 which was introduced in the previous section.

4.3.5 Extensions of SVM by MOP/GP

Considering both the slack variables for misclassified data points (i.e., exterior deviations) and the *surplus variables* for correctly classified data points (i.e., interior deviations), various algorithms of SVM was developed by the authors [101, 163].

4.3 Support Vector Machines for Pattern Classification

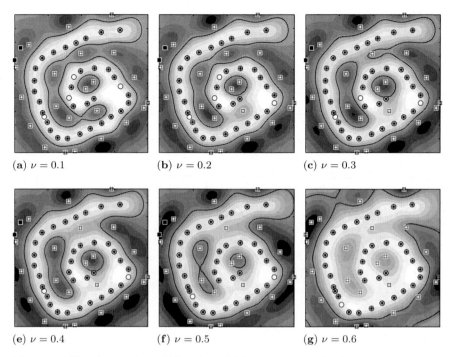

Fig. 4.10 Classification by ν-SVM: the *solid line* represents the discriminant function and the *symbol* ∗ does support vectors

Total Margin Algorithm

In order to minimize the slackness and to maximize the surplus, we consider the following problem:

$$\underset{\boldsymbol{w},b,\boldsymbol{\xi},\boldsymbol{\eta}}{\text{minimize}} \quad \frac{1}{2}\|\boldsymbol{w}\|_2^2 + C_1 \sum_{i=1}^{\ell} \xi_i - C_2 \sum_{i=1}^{\ell} \eta_i \qquad (\text{SVM}_{total})_P$$

$$\text{subject to} \quad y_i\left(\boldsymbol{w}^T \boldsymbol{z}_i + b\right) \geq 1 - \xi_i + \eta_i, \ i = 1,\ldots,\ell,$$

$$\xi_i \geq 0, \ \eta_i \geq 0, \ i = 1,\ldots,\ell,$$

where C_1 and C_2 are chosen in such a way that $C_1 > C_2$ which ensures that at least one of ξ_i and η_i becomes zero.

The Lagrangian function for the problem $(\mathrm{SVM}_{total})_P$ is

$$L(\boldsymbol{w}, b, \boldsymbol{\xi}, \boldsymbol{\eta}, \boldsymbol{\alpha}, \boldsymbol{\beta}, \boldsymbol{\gamma}) = \frac{1}{2}\|\boldsymbol{w}\|_2^2 + C_1 \sum_{i=1}^{\ell} \xi_i - C_2 \sum_{i=1}^{\ell} \eta_i$$

$$- \sum_{i=1}^{\ell} \alpha_i \left(y_i \left(\boldsymbol{w}^T \boldsymbol{z}_i + b \right) - 1 + \xi_i - \eta_i \right)$$

$$- \sum_{i=1}^{\ell} \beta_i \xi_i - \sum_{i=1}^{\ell} \gamma_i \eta_i,$$

where $\alpha_i \geq 0,\ \beta_i \geq 0$ and $\gamma_i \geq 0$.

Differentiating the Lagrangian function with respect to $\boldsymbol{w}$, b, $\boldsymbol{\xi}$ and $\boldsymbol{\eta}$ yields the following conditions:

$$\frac{\partial L(\boldsymbol{w}, b, \boldsymbol{\xi}, \boldsymbol{\eta}, \boldsymbol{\alpha}, \boldsymbol{\beta}, \boldsymbol{\gamma})}{\partial \boldsymbol{w}} = \boldsymbol{w} - \sum_{i=1}^{\ell} \alpha_i y_i \boldsymbol{z}_i = \boldsymbol{0},$$

$$\frac{\partial L(\boldsymbol{w}, b, \boldsymbol{\xi}, \boldsymbol{\eta}, \boldsymbol{\alpha}, \boldsymbol{\beta}, \boldsymbol{\gamma})}{\partial \xi_i} = C_1 - \alpha_i - \beta_i = 0,$$

$$\frac{\partial L(\boldsymbol{w}, b, \boldsymbol{\xi}, \boldsymbol{\eta}, \boldsymbol{\alpha}, \boldsymbol{\beta}, \boldsymbol{\gamma})}{\partial \eta_i} = -C_2 + \alpha_i - \gamma_i = 0,$$

$$\frac{\partial L(\boldsymbol{w}, b, \boldsymbol{\xi}, \boldsymbol{\eta}, \boldsymbol{\alpha}, \boldsymbol{\beta}, \boldsymbol{\gamma})}{\partial b} = \sum_{i=1}^{\ell} \alpha_i y_i = 0.$$

Substituting the above stationary conditions into the Lagrangian function L and using kernel representation, the following dual problem can be obtained:

$$\begin{aligned} \underset{\boldsymbol{\alpha}}{\text{maximize}} \quad & \sum_{i=1}^{\ell} \alpha_i - \frac{1}{2} \sum_{i,j=1}^{\ell} y_i y_j \alpha_i \alpha_j K\left(\boldsymbol{x}_i, \boldsymbol{x}_j\right) \qquad (\mathrm{SVM}_{total}) \\ \text{subject to} \quad & \sum_{i=1}^{\ell} y_i \alpha_i = 0, \\ & C_2 \leq \alpha_i \leq C_1, \ i = 1, \dots, \ell. \end{aligned}$$

Let $\boldsymbol{\alpha}^*$ be the optimal solution to the problem (SVM_{total}). Then, the discriminant function can be written by

$$f(\boldsymbol{x}) = \sum_{i=1}^{\ell} \alpha_i^* y_i K\left(\boldsymbol{x}, \boldsymbol{x}_i\right) + b^*.$$

The offset (or bias) b^* is given as follows: Let n_+ be the number of $\boldsymbol{x}_j$ with $C_2 < \alpha_j^* < C_1$ and $y_j = +1$, and let n_- be the number of $\boldsymbol{x}_j$ with $C_2 < \alpha_j^* < C_1$ and $y_j = -1$, respectively. From the Karush–Kuhn–Tucker

4.3 Support Vector Machines for Pattern Classification

complementarity conditions, if $C_2 < \alpha_j^* < C_1$, then $\beta_j > 0$ and $\gamma_j > 0$. This implies that $\xi_j = \eta_j = 0$. Then,

$$b^* = \frac{1}{n_+ + n_-} \left((n_+ - n_-) - \sum_{j=1}^{n_+ + n_-} \sum_{i=1}^{\ell} y_i \alpha_i^* K\left(\boldsymbol{x}_i, \boldsymbol{x}_j\right) \right).$$

μ-SVM

Introducing a new variable σ which represents the maximal distance between the separating hyperplane and misclassified data points, we formulate the problem $(\mu\text{-SVM})_P$ to minimize the worst slackness and to maximize the sum of surplus. This is a reverse form of ν-SVM, as can be seen from the formulation:

$$\begin{aligned} \underset{\boldsymbol{w},b,\sigma,\boldsymbol{\eta}}{\text{minimize}} \quad & \frac{1}{2}\|\boldsymbol{w}\|_2^2 + \mu\sigma - \frac{1}{\ell}\sum_{i=1}^{\ell}\eta_i \qquad\qquad (\mu\text{-SVM})_P \\ \text{subject to} \quad & y_i\left(\boldsymbol{w}^T\boldsymbol{z}_i + b\right) \geqq \eta_i - \sigma, \ \ i = 1,\ldots,\ell, \\ & \sigma \geqq 0, \ \ \eta_i \geqq 0, \ \ i = 1,\ldots,\ell, \end{aligned}$$

where μ is a parameter which reflects the trade-off between σ and the sum of η_i.

The Lagrangian function for the problem $(\mu\text{-SVM})_P$ is

$$L(\boldsymbol{w},b,\boldsymbol{\eta},\sigma,\boldsymbol{\alpha},\boldsymbol{\beta},\gamma) = \frac{1}{2}\|\boldsymbol{w}\|_2^2 + \mu\sigma - \frac{1}{\ell}\sum_{i=1}^{\ell}\eta_i$$

$$- \sum_{i=1}^{\ell}\alpha_i\left(y_i\left(\boldsymbol{w}^T\boldsymbol{z}_i + b\right) - \eta_i + \sigma\right) - \sum_{i=1}^{\ell}\beta_i\eta_i - \gamma\sigma,$$

where $\alpha_i \geqq 0, \ \beta_i \geqq 0$ and $\gamma \geqq 0$.

Differentiating the Lagrangian function with respect to $\boldsymbol{w}$, b, $\boldsymbol{\eta}$ and σ yields the following conditions:

$$\frac{\partial L(\boldsymbol{w},b,\boldsymbol{\eta},\sigma,\boldsymbol{\alpha},\boldsymbol{\beta},\gamma)}{\partial \boldsymbol{w}} = \boldsymbol{w} - \sum_{i=1}^{\ell}\alpha_i y_i \boldsymbol{z}_i = \boldsymbol{0},$$

$$\frac{\partial L(\boldsymbol{w},b,\boldsymbol{\eta},\sigma,\boldsymbol{\alpha},\boldsymbol{\beta},\gamma)}{\partial \eta_i} = -\frac{1}{\ell} + \alpha_i - \beta_i = 0,$$

$$\frac{\partial L(\boldsymbol{w},b,\boldsymbol{\eta},\sigma,\boldsymbol{\alpha},\boldsymbol{\beta},\gamma)}{\partial \sigma} = \mu - \sum_{i=1}^{\ell}\alpha_i - \gamma = 0,$$

$$\frac{\partial L(\boldsymbol{w},b,\boldsymbol{\eta},\sigma,\boldsymbol{\alpha},\boldsymbol{\beta},\gamma)}{\partial b} = \sum_{i=1}^{\ell}\alpha_i y_i = 0.$$

Substituting the above stationary conditions into the Lagrangian function L, we obtain the following dual optimization problem:

$$\underset{\boldsymbol{\alpha}}{\text{maximize}} \qquad -\frac{1}{2} \sum_{i,j=1}^{\ell} \alpha_i \alpha_j y_i y_j K\left(\boldsymbol{x}_i, \boldsymbol{x}_j\right) \qquad (\mu\text{-SVM})$$

$$\text{subject to} \qquad \sum_{i=1}^{\ell} \alpha_i y_i = 0,$$

$$\sum_{i=1}^{\ell} \alpha_i \leqq \mu,$$

$$\alpha_i \geqq \frac{1}{\ell}, \quad i = 1, \dots, \ell.$$

Let $\boldsymbol{\alpha}^*$ be the optimal solution to the problem (μ-SVM). To compute the offset b^*, we take the set $\mathcal{A}$ of $\boldsymbol{x}_j$ which is the same size n_A with $\frac{1}{\ell} < \alpha_j^*$. From the Karush–Kuhn–Tucker complementarity conditions, if $\frac{1}{\ell} < \alpha_j^*$, then $\beta_j > 0$ which implies $\eta_j = 0$. Thus,

$$b^* = -\frac{1}{2n_A} \sum_{\boldsymbol{x}_j \in \mathcal{A}} \sum_{i=1}^{\ell} \alpha_i^* y_i K\left(\boldsymbol{x}_i, \boldsymbol{x}_j\right).$$

μ-ν-SVM

Applying SVM_{total} and μ-SVM, all training data become support vectors due to the second constraint of the problem (SVM_{total}) and the third constraint of the problem (μ-SVM). In other words, the algorithms (SVM_{total}) and (μ-SVM) lack in the sparsity of support vectors.

In order to overcome this problem in (SVM_{total}) and (μ-SVM), minimizing the worst slackness and maximizing the worst surplus yields the following formulation, which combines the ideas of ν-SVM and μ-SVM:

$$\underset{\boldsymbol{w},b,\rho,\sigma}{\text{minimize}} \qquad \frac{1}{2}\|\boldsymbol{w}\|_2^2 - \nu\rho + \mu\sigma \qquad (\mu\text{-}\nu\text{-SVM})_P$$

$$\text{subject to} \qquad y_i\left(\boldsymbol{w}^T \boldsymbol{z}_i + b\right) \geqq \rho - \sigma, \quad i = 1, \dots, \ell,$$

$$\rho \geqq 0, \quad \sigma \geqq 0,$$

where ν and μ are the trade-off parameters between the norm of $\boldsymbol{w}$, and ρ and σ ($\mu > \nu$).

The Lagrangian function to the problem (μ-ν-SVM)$_P$ is

$$L(\boldsymbol{w}, b, \rho, \sigma, \boldsymbol{\alpha}, \beta, \gamma) = \frac{1}{2}\|\boldsymbol{w}\|_2^2 - \nu\rho + \mu\sigma$$

$$-\sum_{i=1}^{\ell} \alpha_i \left(y_i \left(\boldsymbol{w}^T \boldsymbol{z}_i + b\right) - \rho + \sigma\right) - \beta\rho - \gamma\sigma,$$

where $\alpha_i \geqq 0$, $\beta \geqq 0$ and $\gamma \geqq 0$.

4.3 Support Vector Machines for Pattern Classification

Differentiating Lagrangian function with respect to $\boldsymbol{w}$, b, ρ and σ yields the four conditions

$$\frac{\partial L(\boldsymbol{w}, b, \rho, \sigma, \boldsymbol{\alpha}, \beta, \gamma)}{\partial \boldsymbol{w}} = \boldsymbol{w} - \sum_{i=1}^{\ell} \alpha_i y_i \boldsymbol{z}_i = \boldsymbol{0},$$

$$\frac{\partial L(\boldsymbol{w}, b, \rho, \sigma, \boldsymbol{\alpha}, \beta, \gamma)}{\partial \rho} = -\nu + \sum_{i=1}^{\ell} \alpha_i - \beta = 0,$$

$$\frac{\partial L(\boldsymbol{w}, b, \rho, \sigma, \boldsymbol{\alpha}, \beta, \gamma)}{\partial \sigma} = \mu - \sum_{i=1}^{\ell} \alpha_i - \gamma = 0,$$

$$\frac{\partial L(\boldsymbol{w}, b, \rho, \sigma, \boldsymbol{\alpha}, \beta, \gamma)}{\partial b} = \sum_{i=1}^{\ell} \alpha_i y_i = 0.$$

Substituting the above stationary conditions into the Lagrangian function L and using kernel representation, we obtain the following dual optimization problem:

$$\begin{aligned}
\underset{\boldsymbol{\alpha}}{\text{maximize}} \quad & -\frac{1}{2} \sum_{i,j=1}^{\ell} \alpha_i \alpha_j y_i y_j K(\boldsymbol{x}_i, \boldsymbol{x}_j) \quad (\mu\text{-}\nu\text{-SVM}) \\
\text{subject to} \quad & \sum_{i=1}^{\ell} \alpha_i y_i = 0, \\
& \nu \leqq \sum_{i=1}^{\ell} \alpha_i \leqq \mu, \\
& \alpha_i \geqq 0, \quad i = 1, \ldots, \ell.
\end{aligned}$$

Letting $\boldsymbol{\alpha}^*$ be the optimal solution to the problem (μ-ν-SVM), the offset b^* can be chosen easily for any i satisfying $\alpha_i^* > 0$. Otherwise, b^* can be obtained by the similar way with the decision of the b^* in the other algorithms.

Figure 4.11 shows the results by μ-ν-SVM for Example 4.2 which was used in the previous section.

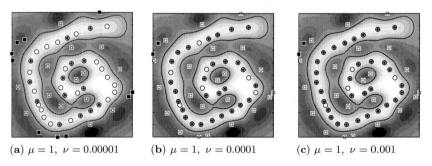

(a) $\mu = 1$, $\nu = 0.00001$ (b) $\mu = 1$, $\nu = 0.0001$ (c) $\mu = 1$, $\nu = 0.001$

Fig. 4.11 Classification by μ-ν-SVM: the *solid line* represents the discriminant function and the *symbol* ∗ does support vectors

Remark 4.4. Note that at least either ρ or σ vanishes at the solution in the problem $(\mu\text{-}\nu\text{-SVM})_P$. Indeed, if both of the solutions $\hat{\rho}$ and $\hat{\sigma}$ are positive, we have $\hat{\rho} - \delta > 0$ and $\hat{\sigma} - \delta > 0$ for a sufficiently small $\delta > 0$ satisfying the constraints of $(\mu\text{-}\nu\text{-SVM})_P$ and yielding a less value of objective function due to $\mu > \nu$, which leads to a contradiction. The fact $\hat{\sigma} = 0$ means that we have a perfect classification for the training data. In this event, therefore, $\mu\text{-}\nu\text{-SVM}$ becomes equivalent to the hard margin SVM. Using Gaussian kernels, we have $\hat{\sigma} = 0$ in many cases. However, we may have $\hat{\sigma} \neq 0$ (hence, $\hat{\rho} = 0$) with another kernel yielding a finite-dimensional feature space, e.g., polynomial kernel. In fact, in the following section, we have $\hat{\sigma} \neq 0$ in Case 2 of Table 4.7 for MONK's problem using the quadratic polynomial kernel.

4.3.6 Comparison of Experimental Results

In order to investigate the performance of several SVM algorithms, we compare the results for MONK's problem which can be downloaded from http://www.ics.uci.edu/mlearn/MLSummary.html:

1. *Case 1*
   (a) Training: 124 instances ($\mathcal{A}$: 62 instances, $\mathcal{B}$: 62 instances)
   (b) Test: 432 instances ($\mathcal{A}$: 216 instances, $\mathcal{B}$: 216 instances)
2. *Case 2*
   (a) Training: 169 instances ($\mathcal{A}$: 64 instances, $\mathcal{B}$: 105 instances)
   (b) Test: 432 instances ($\mathcal{A}$: 142 instances, $\mathcal{B}$: 290 instances)
3. *Case 3*
   (a) Training: 122 instances ($\mathcal{A}$: 60 instances, $\mathcal{B}$: 62 instances)
   (b) Test: 432 instances ($\mathcal{A}$: 228 instances, $\mathcal{B}$: 204 instances)

In the numerical experiments, QP solver of MATLAB was used for solving QP problems in SVM formulations. Although the Gaussian kernel yielded better classification, we used here the quadratic polynomial kernel in order to show the result of Remark 4.4.

The data were normalized for each sample $x_i = (x_{1i}, \ldots, x_{ni})^T$ such that

$$\tilde{x}_{ki} = \frac{x_{ki} - m_k}{s_k}, \quad k = 1, \ldots, n,$$

where m_k and s_k are the mean value and the standard deviation of kth component of given training input data $\{x_1, \ldots, x_\ell\}$, respectively. For the parameters in applying MOP/GP$_1$ model, we set $h_1 = \cdots = h_\ell = C_1$ and

4.3 Support Vector Machines for Pattern Classification

$k_1 = \cdots = k_\ell = C_2$. In the computations, we use both the training data set and the test data set as in the benchmark of the web site. Tables 4.1–4.7 show the classification rates by using the existing algorithms MOP/GP$_1$, SVM$_{hard}$, SVM$_{soft}$ and ν-SVM with the MOP/GP approach algorithms SVM$_{total}$, μ-SVM and μ-ν-SVM.

Throughout the numerical experiments, it has been observed that even though the result depends on the value of parameters, the family of SVM using MOP/GP approach such as ν-SVM, SVM$_{total}$, μ-SVM and μ-ν-SVM shows a relatively good performance in comparison with the simple SVM$_{soft}$. Sometimes unbalanced data sets cause a difficulty in predicting the category with fewer samples. From the results for Case 2, it is seen that the classification ability for the class with fewer samples is much sensitive to the value of C in SVM$_{soft}$. In other words, we have to select the appropriate value of C in SVM$_{soft}$ carefully in order to attain some reasonable classification rate for unbalanced data sets. SVM$_{total}$ and μ-ν-SVM, however, have advantage over SVM$_{soft}$ in classification rate of the class with fewer elements. In addition, the data set of MONK seems not to be linearly separated. In this example, therefore, the SVMs using MOP/GP approach show much better performance than the mere GP (for more results and details, see [101]).

Table 4.1 Classification rate by GP

	C_1	1			10			100			
	C_2	0.001	0.01	0.1	0.01	0.1	1	0.1	1	10	average
Training	Case 1	73.39	73.39	73.39	73.39	73.39	71.77	73.39	73.39	73.39	73.21
	Case 2	63.31	63.31	63.31	63.31	63.31	63.91	63.31	63.31	65.09	63.57
	$\mathcal{A}$	53.13	53.13	45.31	51.56	51.56	48.44	51.56	51.56	45.31	50.17
	$\mathcal{B}$	69.52	69.52	74.29	70.48	70.48	73.33	70.48	70.48	77.14	71.75
	Case 3	88.52	88.52	88.52	88.52	88.52	88.52	88.52	88.52	88.52	88.52
Test	Case 1	66.67	66.67	66.67	66.67	66.67	65.97	66.67	66.67	66.67	66.59
	Case 2	58.33	58.33	58.33	59.03	59.03	59.26	59.03	59.03	61.11	59.05
	$\mathcal{A}$	39.44	39.44	35.92	40.14	40.14	37.32	40.14	40.14	35.92	38.73
	$\mathcal{B}$	67.59	67.59	70.69	68.28	68.28	70.00	68.28	68.28	73.45	69.16
	Case 3	88.89	88.89	88.89	88.89	88.89	88.89	88.89	88.89	88.89	88.89

Table 4.2 Classification rate by SVM$_{hard}$

		Training			Test			Support vectors
			$\mathcal{A}$	$\mathcal{B}$		$\mathcal{A}$	$\mathcal{B}$	
Case 1	100.00	100.00	100.00	83.80	83.80	83.80	20.16
Case 2	–	–	–	–	–	–	-
Case 3	100.00	100.00	100.00	90.28	91.67	88.73	18.85
Average	100.00	100.00	100.00	87.04	87.73	86.26	19.51
STD	0.00	0.00	0.00	3.24	3.94	2.46	0.65

Table 4.3 Classification rate by SVM$_{soft}$

	C	Training			Test			Support vectors
			$\mathcal{A}$	$\mathcal{B}$		$\mathcal{A}$	$\mathcal{B}$	
Case 1	0.1	89.52	85.48	93.55	81.48	79.63	83.33	46.77
	0.5	92.74	88.71	96.77	84.49	82.87	86.11	34.68
	1	95.16	91.94	98.39	84.26	79.17	89.35	33.87
	5	98.39	100.00	96.77	85.19	86.57	83.80	25.00
	10	99.19	100.00	98.39	84.49	84.26	84.72	23.39
	50	100.00	100.00	100.00	83.80	83.80	83.80	20.16
	100	100.00	100.00	100.00	83.80	83.80	83.80	20.16
Case 2	0.1	80.47	70.31	86.67	71.53	61.27	76.55	66.27
	0.5	81.66	78.13	83.81	74.31	71.13	75.86	53.25
	1	85.80	85.94	85.71	77.31	76.06	77.93	50.89
	5	84.02	81.25	85.71	76.62	72.54	78.62	46.15
	10	84.02	81.25	85.71	76.62	72.54	78.62	46.15
	50	85.21	82.81	86.67	78.01	74.65	79.66	44.97
	100	85.21	82.81	86.67	78.01	74.65	79.66	44.97
Case 3	0.1	97.54	100.00	95.16	93.75	93.86	93.63	34.43
	0.5	98.36	100.00	96.77	92.59	91.67	93.63	25.41
	1	99.18	100.00	98.39	92.36	91.67	93.14	22.95
	5	99.18	100.00	98.39	91.67	90.35	93.14	20.49
	10	100.00	100.00	100.00	89.81	91.23	88.24	19.67
	50	100.00	100.00	100.00	90.28	91.67	88.73	18.85
	100	100.00	100.00	100.00	90.28	91.67	88.73	18.85
Average		93.13	91.84	93.98	83.84	82.14	84.81	34.16
STD		7.15	9.40	5.99	6.52	8.69	5.73	13.83

4.4 Support Vector Machines for Regression

Support vector machine was extended to regression by introducing the ε-*insensitive loss function* by Vapnik [153].

Denote the given training data set by $(\boldsymbol{x}_i, y_i)$, $i = 1, \ldots, \ell$. Suppose that the regression function f on the feature space Z is expressed by

$$f(\boldsymbol{z}) = \sum_{i=1}^{\ell} w_i \boldsymbol{z}_i + b,$$

and the linear ε-insensitive loss function (Fig. 4.12) is defined by

$$L^{\varepsilon}(\boldsymbol{z}, y, f) = |y - f(\boldsymbol{z})|_{\varepsilon} = \max(0, |y - f(\boldsymbol{z})| - \varepsilon).$$

4.4 Support Vector Machines for Regression

Table 4.4 Classification rate by ν-SVM

	ν	Training			Test			Support vectors
			$\mathcal{A}$	$\mathcal{B}$		$\mathcal{A}$	$\mathcal{B}$	
Case 1	0.1	99.19	100.00	98.39	84.49	83.80	85.19	23.39
	0.2	95.97	95.16	96.77	83.80	80.56	87.04	30.65
	0.3	89.52	87.10	91.94	81.02	80.56	81.48	38.71
	0.4	89.52	85.48	93.55	81.48	79.17	83.80	47.58
	0.5	88.71	79.03	98.39	81.48	73.15	89.81	58.87
	0.6	89.52	80.65	98.39	81.02	73.15	88.89	66.13
	0.7	87.90	77.42	98.39	79.17	71.30	87.04	75.81
Case 2	0.1	81.66	73.44	86.67	75.46	68.31	78.97	66.27
	0.2	81.66	73.44	86.67	75.46	68.31	78.97	65.68
	0.3	81.66	73.44	86.67	75.23	68.31	78.62	65.09
	0.4	85.80	84.38	86.67	77.55	75.35	78.62	49.70
	0.5	82.25	78.13	84.76	72.92	67.61	75.52	56.80
	0.6	81.07	70.31	87.62	71.76	61.27	76.90	66.27
	0.7	75.15	56.25	86.67	67.59	43.66	79.31	74.56
Case 3	0.1	99.18	100.00	98.39	92.36	91.67	93.14	23.77
	0.2	96.72	98.33	95.16	92.13	90.79	93.63	31.97
	0.3	96.72	100.00	93.55	95.37	94.30	96.57	40.98
	0.4	95.90	98.33	93.55	94.44	92.98	96.08	49.18
	0.5	95.90	98.33	93.55	94.21	91.67	97.06	55.74
	0.6	94.26	91.67	96.77	90.74	84.21	98.04	68.03
	0.7	94.26	91.67	96.77	89.35	81.58	98.04	74.59
Average		89.64	85.36	92.82	82.72	77.22	86.79	53.80
STD		6.87	11.96	4.84	8.10	12.07	7.61	16.37

For a given insensitivity parameter $\varepsilon > 0$, the degree of regression error should be minimized (Fig. 4.13),

$$\underset{\boldsymbol{w},b,\boldsymbol{\xi},\acute{\boldsymbol{\xi}}}{\text{minimize}} \qquad \frac{1}{2}\|\boldsymbol{w}\|_2^2 + C\left(\frac{1}{\ell}\sum_{i=1}^{\ell}(\xi_i + \acute{\xi}_i)\right) \qquad (C\text{-SVR})_P$$

$$\text{subject to} \qquad (\boldsymbol{w}^T\boldsymbol{z}_i + b) - y_i \leqq \varepsilon + \xi_i, \ \ i = 1,\ldots,\ell,$$

$$y_i - (\boldsymbol{w}^T\boldsymbol{z}_i + b) \leqq \varepsilon + \acute{\xi}_i, \ \ i = 1,\ldots,\ell,$$

$$\xi_i, \ \acute{\xi}_i \geqq 0,$$

where C is a trade-off parameter between the norm of $\boldsymbol{w}$ and ξ_i $(\acute{\xi}_i)$.

Table 4.5 Classification rate by SVM$_{total}$

	C_1	C_2	Training			Test			Support vectors
				$\mathcal{A}$	$\mathcal{B}$		$\mathcal{A}$	$\mathcal{B}$	
Case 1	1	0.001	95.16	91.94	98.39	84.26	79.17	89.35	100.00
		0.01	95.97	95.16	96.77	84.26	81.02	87.50	100.00
		0.1	96.77	96.77	96.77	81.71	84.72	78.70	100.00
	10	0.01	99.19	100.00	98.39	84.49	84.26	84.72	100.00
		0.1	99.19	100.00	98.39	82.64	83.80	81.48	100.00
		1	96.77	96.77	96.77	80.09	82.87	77.31	100.00
	100	0.1	100.00	100.00	100.00	83.56	84.26	82.87	100.00
		1	100.00	100.00	100.00	81.71	82.87	80.56	100.00
		10	96.77	96.77	96.77	79.86	82.41	77.31	100.00
Case 2	1	0.001	85.80	85.94	85.71	77.31	76.06	77.93	100.00
		0.01	85.80	84.38	86.67	77.31	75.35	78.28	100.00
		0.1	83.43	76.56	87.62	78.24	68.31	83.10	100.00
	10	0.01	84.62	81.25	86.67	77.08	72.54	79.31	100.00
		0.1	85.21	82.81	86.67	78.01	74.65	79.66	100.00
		1	86.39	81.25	89.52	79.40	66.90	85.52	100.00
	100	0.1	85.21	82.81	86.67	78.01	74.65	79.66	100.00
		1	85.21	82.81	86.67	78.01	74.65	79.66	100.00
		10	85.80	79.69	89.52	78.70	65.49	85.17	100.00
Case 3	1	0.001	99.18	100.00	98.39	92.13	91.23	93.14	100.00
		0.01	99.18	100.00	98.39	92.59	91.23	94.12	100.00
		0.1	98.36	98.33	98.39	91.20	85.96	97.06	100.00
	10	0.01	100.00	100.00	100.00	90.51	91.67	89.22	100.00
		0.1	100.00	100.00	100.00	91.67	89.47	94.12	100.00
		1	98.36	98.33	98.39	91.20	85.96	97.06	100.00
	100	0.1	100.00	100.00	100.00	91.67	89.47	94.12	100.00
		1	100.00	100.00	100.00	92.82	89.47	96.57	100.00
		10	98.36	98.33	98.39	91.20	85.96	97.06	100.00
Average			94.10	92.96	94.81	84.06	81.27	85.95	100.00
STD			6.38	8.13	5.45	5.79	7.49	6.98	0.00

The dual formulation to the problem $(C\text{-SVR})_P$ is given by

$$\underset{\alpha,\acute{\alpha}}{\text{maximize}} \quad -\frac{1}{2}\sum_{i,j=1}^{\ell} (\acute{\alpha}_i - \alpha_i)(\acute{\alpha}_j - \alpha_j) K(\boldsymbol{x}_i, \boldsymbol{x}_j) \qquad (C\text{-SVR})$$

$$+\sum_{i=1}^{\ell} (\acute{\alpha}_i - \alpha_i) y_i - \varepsilon \sum_{i=1}^{\ell} (\acute{\alpha}_i + \alpha_i)$$

$$\text{subject to} \quad \sum_{i=1}^{\ell} (\acute{\alpha}_i - \alpha_i) = 0,$$

$$0 \leqq \acute{\alpha}_i \leqq \frac{C}{\ell}, \ \ 0 \leqq \alpha_i \leqq \frac{C}{\ell}, \ \ i = 1,\ldots,\ell.$$

Figure 4.14 shows a numerical example by C-SVR.

4.4 Support Vector Machines for Regression

Table 4.6 Classification rate by μ-SVM

	μ	Training			Test			Support vectors
			$\mathcal{A}$	$\mathcal{B}$		$\mathcal{A}$	$\mathcal{B}$	
Case 1	1.5	82.26	85.48	79.03	70.14	75.46	64.81	100.00
	2	83.06	85.48	80.65	72.69	77.78	67.59	100.00
	5	82.26	85.48	79.03	72.92	78.70	67.13	100.00
	10	82.26	87.10	77.42	74.54	80.56	68.52	100.00
	50	84.68	88.71	80.65	76.39	81.48	71.30	100.00
	100	84.68	85.48	83.87	77.08	79.17	75.00	100.00
	1,000	100.00	100.00	100.00	81.25	82.41	80.09	100.00
Case 2	1.5	74.56	76.56	73.33	66.20	64.08	67.24	100.00
	2	75.15	78.13	73.33	65.74	65.49	65.86	100.00
	5	76.33	59.38	86.67	72.22	49.30	83.45	100.00
	10	76.33	60.94	85.71	72.22	50.00	83.10	100.00
	50	77.51	65.63	84.76	72.69	52.82	82.41	100.00
	100	75.15	70.31	78.10	73.15	60.56	79.31	100.00
	1,000	100.00	100.00	100.00	100.00	100.00	100.00	100.00
Case 3	1.5	92.62	88.33	96.77	87.96	80.70	96.08	100.00
	2	92.62	88.33	96.77	88.66	81.58	96.57	100.00
	5	90.98	85.00	96.77	89.12	81.14	98.04	100.00
	10	92.62	88.33	96.77	90.05	82.89	98.04	100.00
	50	92.62	88.33	96.77	90.74	85.09	97.06	100.00
	100	100.00	100.00	100.00	93.52	90.35	97.06	100.00
	1,000	100.00	100.00	100.00	93.52	90.35	97.06	100.00
Average		86.46	84.14	87.92	80.04	75.71	82.65	100.00
STD		8.89	11.74	9.52	9.93	13.32	12.88	0.00

In order to decide the insensitivity parameter ε automatically, Schölkopf and Smola proposed ν-SVR which is formulated by the following [132]:

$$\underset{\boldsymbol{w},b,\varepsilon,\boldsymbol{\xi},\acute{\boldsymbol{\xi}}}{\text{minimize}} \qquad \frac{1}{2}\|\boldsymbol{w}\|_2^2 + C\left(\nu\varepsilon + \frac{1}{\ell}\sum_{i=1}^{\ell}(\xi_i + \acute{\xi}_i)\right) \qquad (\nu\text{-SVR})_P$$

$$\text{subject to} \qquad \left(\boldsymbol{w}^T\boldsymbol{z}_i + b\right) - y_i \leqq \varepsilon + \xi_i, \ i = 1,\ldots,\ell,$$

$$y_i - \left(\boldsymbol{w}^T\boldsymbol{z}_i + b\right) \leqq \varepsilon + \acute{\xi}_i, \ i = 1,\ldots,\ell,$$

$$\varepsilon, \ \xi_i, \ \acute{\xi}_i \geqq 0,$$

where C and $0 < \nu \leqq 1$ are trade-off parameters between the norm of $\boldsymbol{w}$ and ε and ξ_i $(\acute{\xi}_i)$.

Table 4.7 Classification rate by μ-ν-SVM

	μ	ν	Training			Test			Support vectors
				$\mathcal{A}$	$\mathcal{B}$		$\mathcal{A}$	$\mathcal{B}$	
Case 1	1	0.001	100.00	100.00	100.00	83.80	83.80	83.80	20.16
		0.01	100.00	100.00	100.00	83.80	83.80	83.80	20.16
		0.1	100.00	100.00	100.00	83.80	83.80	83.80	20.16
	10	0.01	100.00	100.00	100.00	83.80	83.80	83.80	20.16
		0.1	100.00	100.00	100.00	83.80	83.80	83.80	20.16
		1	100.00	100.00	100.00	83.80	83.80	83.80	20.16
	100	0.1	100.00	100.00	100.00	83.80	83.80	83.80	20.16
		1	100.00	100.00	100.00	83.80	83.80	83.80	20.16
		10	100.00	100.00	100.00	83.80	83.80	83.80	20.16
Case 2	1	0.001	81.66	73.44	86.67	75.46	69.01	78.62	85.80
		0.01	81.66	73.44	86.67	75.00	68.31	78.28	83.43
		0.1	81.66	73.44	86.67	75.46	68.31	78.97	81.07
	10	0.01	81.66	73.44	86.67	75.00	68.31	78.28	83.43
		0.1	81.66	73.44	86.67	75.46	68.31	78.97	81.07
		1	81.66	73.44	86.67	75.46	68.31	78.97	84.62
	100	0.1	81.66	73.44	86.67	75.46	68.31	78.97	81.07
		1	81.66	73.44	86.67	75.46	68.31	78.97	84.62
		10	81.66	73.44	86.67	75.00	67.61	78.62	86.98
Case 3	1	0.001	100.00	100.00	100.00	90.28	91.67	88.73	18.85
		0.01	100.00	100.00	100.00	90.28	91.67	88.73	18.85
		0.1	100.00	100.00	100.00	90.28	91.67	88.73	18.85
	10	0.01	100.00	100.00	100.00	90.28	91.67	88.73	18.85
		0.1	100.00	100.00	100.00	90.28	91.67	88.73	18.85
		1	100.00	100.00	100.00	90.28	91.67	88.73	18.85
	100	0.1	100.00	100.00	100.00	90.28	91.67	88.73	18.85
		1	100.00	100.00	100.00	90.28	91.67	88.73	18.85
		10	100.00	100.00	100.00	90.28	91.67	88.73	18.85
Average			93.89	91.15	95.56	83.13	81.26	83.75	40.86
STD			8.65	12.52	6.29	6.13	9.70	4.08	30.22

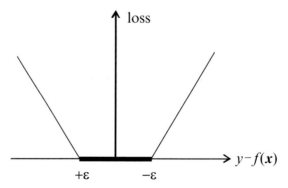

Fig. 4.12 ε-insensitive loss function

4.4 Support Vector Machines for Regression

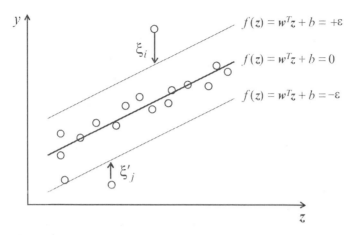

Fig. 4.13 Support vector regression

The dual formulation to the problem $(\nu\text{-SVR})_P$ is given by

$$\underset{\alpha,\acute{\alpha}}{\text{maximize}} \quad -\frac{1}{2}\sum_{i,j=1}^{\ell}(\acute{\alpha}_i - \alpha_i)(\acute{\alpha}_j - \alpha_j)K(\boldsymbol{x}_i, \boldsymbol{x}_j) \quad (\nu\text{-SVR})$$

$$+ \sum_{i=1}^{\ell}(\acute{\alpha}_i - \alpha_i)y_i$$

subject to
$$\sum_{i=1}^{\ell}(\acute{\alpha}_i - \alpha_i) = 0,$$

$$\sum_{i=1}^{\ell}(\acute{\alpha}_i + \alpha_i) \leqq C\nu,$$

$$0 \leqq \acute{\alpha}_i \leqq \frac{C}{\ell}, \ 0 \leqq \alpha_i \leqq \frac{C}{\ell}, \ i = 1,\ldots,\ell.$$

Figure 4.15 shows a numerical examply by ν-SVR.

In a similar fashion to classification, $(C\text{-SVR})$ can be extended to $(\mu\text{-SVR})$ as follows.

For a given insensitivity parameter ε,

$$\underset{\boldsymbol{w},b,\xi,\acute{\xi}}{\text{minimize}} \quad \frac{1}{2}\|\boldsymbol{w}\|_2^2 + \mu(\xi + \acute{\xi}) \quad (\mu\text{-SVR})_P$$

subject to
$$(\boldsymbol{w}^T\boldsymbol{z}_i + b) - y_i \leqq \varepsilon + \xi, \ i = 1,\ldots,\ell,$$

$$y_i - (\boldsymbol{w}^T\boldsymbol{z}_i + b) \leqq \varepsilon + \acute{\xi}, \ i = 1,\ldots,\ell,$$

$$\varepsilon, \ \xi, \ \acute{\xi} \geqq 0,$$

where μ is a trade-off parameter between the norm of $\boldsymbol{w}$ and ξ ($\acute{\xi}$).

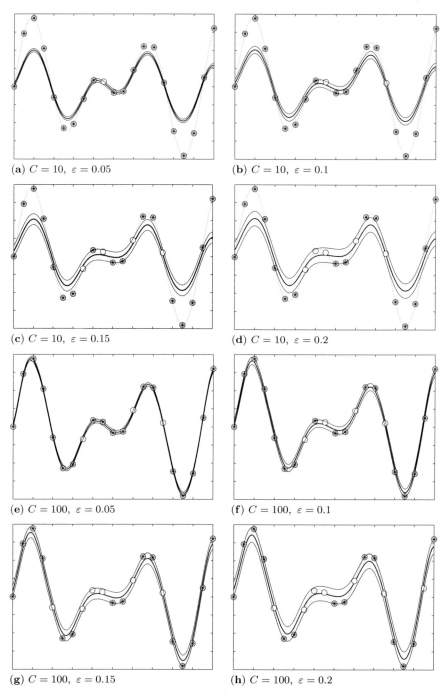

Fig. 4.14 Regression functions by C-SVR: the *dotted line* is the true function, the *thick line* a predicted function $y = f(x)$, and *solid lines* are $y = f(x) \pm \varepsilon$. The *symbol* $*$ shows support vectors

4.4 Support Vector Machines for Regression

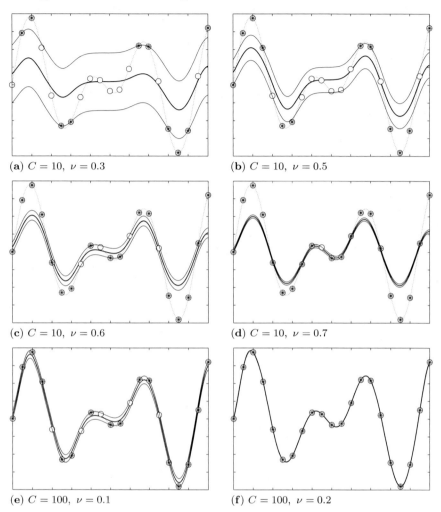

Fig. 4.15 Regression functions by ν-SVR: the *dotted line* is the true function, the *thick line* a predicted function $y = f(x)$, and *solid lines* are $y = f(x) \pm \varepsilon$. The *symbol* $*$ shows support vectors

The Lagrangian function for the problem $(\mu\text{-SVR})_P$ is

$$L(\boldsymbol{w}, b, \xi, \acute{\xi}, \boldsymbol{\alpha}, \boldsymbol{\acute{\alpha}}, \beta, \acute{\beta})$$
$$= \frac{1}{2}\|\boldsymbol{w}\|_2^2 + \mu(\xi + \acute{\xi}) - \sum_{i=1}^{\ell} \alpha_i \left(\varepsilon + \xi + y_i - \boldsymbol{w}^T \boldsymbol{z}_i - b\right)$$
$$- \sum_{i=1}^{\ell} \acute{\alpha}_i \left(\varepsilon + \acute{\xi} - y_i + \boldsymbol{w}^T \boldsymbol{z}_i + b\right) - \beta \xi - \acute{\beta}\acute{\xi},$$

where $\alpha \geqq 0$, $\acute{\alpha} \geqq 0$, $\beta \geqq 0$ and $\acute{\beta} \geqq 0$.

Differentiating the Lagrangian function with respect to w, b and $\xi^{(')}$ yields the following conditions:

$$\frac{\partial L}{\partial w} = w + \sum_{i=1}^{\ell} (\alpha_i - \acute{\alpha}_i) z_i = \mathbf{0},$$

$$\frac{\partial L}{\partial b} = \sum_{i=1}^{\ell} (\alpha_i - \acute{\alpha}_i) = 0,$$

$$\frac{\partial L}{\partial \xi^{(')}} = -\mu + \sum_{i=1}^{\ell} \alpha_i^{(')} - \beta^{(')} = 0.$$

Substituting the above stationary conditions into the Lagrangian function L and using kernel representation, the dual formulation to the problem $(\mu\text{-SVR})_P$ is given by

$$\underset{\alpha,\acute{\alpha}}{\text{maximize}} \quad -\frac{1}{2} \sum_{i,j=1}^{\ell} (\acute{\alpha}_i - \alpha_i)(\acute{\alpha}_j - \alpha_j) K(x_i, x_j) \qquad (\mu\text{-SVR})$$

$$+ \sum_{i=1}^{\ell} (\acute{\alpha}_i - \alpha_i) y_i - \varepsilon \sum_{i=1}^{\ell} (\acute{\alpha}_i + \alpha_i)$$

$$\text{subject to} \quad \sum_{i=1}^{\ell} (\acute{\alpha}_i - \alpha_i) = 0,$$

$$\sum_{i=1}^{\ell} \acute{\alpha}_i \leqq \mu, \ \sum_{i=1}^{\ell} \alpha_i \leqq \mu,$$

$$\acute{\alpha}_i \geqq 0, \ \alpha_i \geqq 0, \ i = 1, \dots, \ell.$$

Figure 4.16 shows a numerical example by μ-SVR.

Remark 4.5. Combining μ-SVR and ν-SVR, we can derive another formulation which may be defined as μ-ν-SVR:

$$\underset{w,b,\varepsilon,\xi,\acute{\xi}}{\text{minimize}} \quad \frac{1}{2} \|w\|_2^2 + \nu\varepsilon + \mu(\xi + \acute{\xi}) \qquad (\mu\text{-}\nu\text{-SVR})_P$$

$$\text{subject to} \quad (w^T z_i + b) - y_i \leqq \varepsilon + \xi, \ i = 1, \dots, \ell,$$

$$y_i - (w^T z_i + b) \leqq \varepsilon + \acute{\xi}, \ i = 1, \dots, \ell,$$

$$\varepsilon, \ \xi, \ \acute{\xi} \geqq 0,$$

where ξ $(\acute{\xi})$ denotes the maximum outer deviation from the ε band, and ν and μ are trade-off parameters between the norm of w and ε and ξ $(\acute{\xi})$, respectively.

In this formulation, however, at least either ε or ξ (or $\acute{\xi}$) vanishes at the solution according to $\nu \geqq 2\mu$ or $\nu \leqq 2\mu$. Therefore, μ-ν-SVR may

4.4 Support Vector Machines for Regression

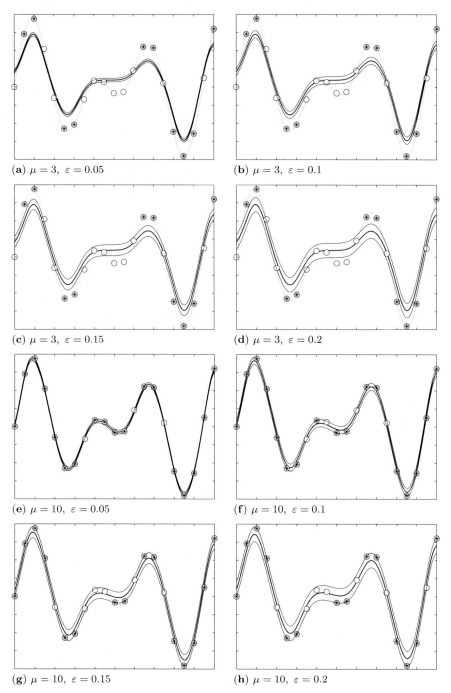

Fig. 4.16 Regression functions by μ-SVR: the *dotted line* is the true function, the *thick line* a predicted function $y = f(x)$, and *solid lines* are $y = f(x) \pm \varepsilon$. The *symbol* $*$ shows support vectors

be reduced to the following formulation simply called ν_ε-SVR (or similarly μ_ξ-SVR replacing ε by ξ ($\acute{\xi}$) and $\nu\varepsilon$ by $\mu(\xi + \acute{\xi})$):

$$\underset{\boldsymbol{w},b,\varepsilon}{\text{minimize}} \qquad \frac{1}{2}\|\boldsymbol{w}\|_2^2 + \nu\varepsilon \qquad\qquad (\nu_\varepsilon - \text{SVR})_P$$

$$\text{subject to} \qquad \left(\boldsymbol{w}^T\boldsymbol{z}_i + b\right) - y_i \leqq \varepsilon, \quad i = 1,\dots,\ell,$$

$$y_i - \left(\boldsymbol{w}^T\boldsymbol{z}_i + b\right) \leqq \varepsilon, \quad i = 1,\dots,\ell,$$

$$\varepsilon \geqq 0,$$

where ν is a trade-off parameter between the norm of $\boldsymbol{w}$ and ε. This formulation can be regarded as a kind of Tchebyshev approximation in the feature space.

The dual formulation of $(\nu_\varepsilon\text{-SVR})_P$ is given by

$$\underset{\alpha,\acute{\alpha}}{\text{maximize}} \qquad -\frac{1}{2}\sum_{i,j=1}^{\ell} (\acute{\alpha}_i - \alpha_i)(\acute{\alpha}_j - \alpha_j) K\left(\boldsymbol{x}_i, \boldsymbol{x}_j\right) \qquad (\nu_\varepsilon\text{-SVR})$$

$$+ \sum_{i=1}^{\ell} (\acute{\alpha}_i - \alpha_i) y_i$$

$$\text{subject to} \qquad \sum_{i=1}^{\ell} (\acute{\alpha}_i - \alpha_i) = 0,$$

$$\sum_{i=1}^{\ell} (\acute{\alpha}_i + \alpha_i) \leqq \nu,$$

$$\acute{\alpha}_i \geqq 0, \quad \alpha_i \geqq 0, \quad i = 1,\dots,\ell.$$

Figure 4.17 shows a numerical example by ν_ε-SVR.

Here, in order to compare the performance of several SVR algorithms described above, consider a high-dimensional chaotic system generated by Mackey Glass delay differential equation [82] as follows:

$$\frac{dx(t)}{dt} = -\beta x(t) + \alpha \frac{x(t-\tau)}{1 + (x(t-\tau))^{10}} \qquad (4.8)$$

with $\alpha = 0.2$, $\beta = 0.1$ and the delay $\tau = 17$. $x(t)$ is predicted from $(x(t-6), x(t-12), x(t-18), x(t-24), x(t-30), x(t-36)$, and the parameters are given as below:

- $x(0) = 1.2$
- Training time $t = 36,\dots,335$
- Test time $t = 336,\dots,365$
- $C - \text{SVR}$: $C = 100$ and $\varepsilon = 0.001$
- ν-SVR: $C = 100$ and $\nu = 0.2$
- μ-SVR: $\mu = 20$ and $\varepsilon = 0.001$

4.4 Support Vector Machines for Regression

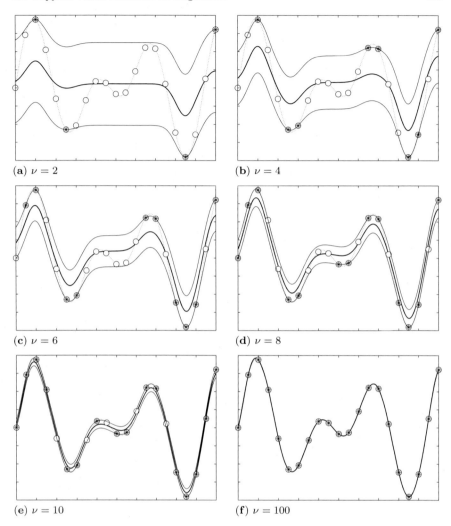

Fig. 4.17 Regression functions by ν_ε-SVR: the *dotted line* is the true function, the *thick line* a predicted function $y = f(x)$, and *solid lines* are $y = f(x) \pm \varepsilon$. The *symbol* $*$ shows support vectors

- ν_ε-SVR: $\nu = 20$
- Added noises follow $N(0, 0.25)$

We simulate two cases of training data with/without noise. Table 4.8 shows comparison results for several kinds of SVRs (in cases with random normal noise, experiments were made ten times for noises generated randomly). One may see that among the methods, ν_ε-SVR provides the least number of support vectors while keeping a reasonable error rate.

Table 4.8 Comparison of the results (unit: %)

			C-SVR		ν-SVR		μ-SVR		ν_ε-SVR	
			Train	Test	Train	Test	Train	Test	Train	Test
No noise	SV rate		55.33		31.67		31.33		19.33	
	Error rate		0.12	3.79	0.20	3.83	0.11	3.73	0.22	3.74
Adding noise	SV rate	Mean	97.97		30.77		24.27		16.27	
		STD	0.79		1.19		3.20		1.81	
	Error rate	Mean	1.87	4.78	2.53	4.75	2.65	5.29	3.01	4.67
		STD	0.09	1.21	0.13	1.06	0.24	1.00	0.26	0.90

As shown in the results, it can be observed that μ-SVR and ν_ε-SVR provide less number of support vectors while keeping a reasonable error rate, compared with C-SVR and ν-SVR, i.e., μ-SVR and ν_ε-SVR is promising for sparse approximation. This means the computation in μ-SVR and ν_ε-SVR is less expensive. The fact that μ-SVR and ν_ε-SVR yield good function approximation with reasonable accuracy and with less support vectors, is important in practice in engineering design. For some kinds of engineering design problems, approximation functions should be realized on the basis of as few data points as possible. Therefore, μ-SVR and ν_ε-SVR will be shown as a useful tool of function approximation in Chap. 5.

4.5 Combining Predetermined Model and SVR/RBFN

Consider a model consisting of a predetermined model and SVR/RBF. Namely, suppose that an approximate function is given by

$$\hat{f}(\boldsymbol{x}) = g(\boldsymbol{x}) + h(\boldsymbol{x}), \tag{4.9}$$

where $g(\boldsymbol{x})$ is a *predetermined model*, for example linear polynomial model and $h(\boldsymbol{x})$ is a SVR/RBFN model which approximates the deviation from the predetermined model.

For example, consider the following problem:

$$f(x) = 1 - x - 0.1\sin(9\pi x). \tag{4.10}$$

For this problem, we show the result by a linear polynomial model in Fig. 4.18. The linear polynomial model can predict well for the linear factor globally, but not approximate the periodical factor locally.

4.5 Combining Predetermined Model and SVR/RBFN

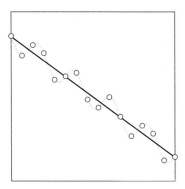

Fig. 4.18 Linear polynomial regression

On the other hand, SVR with Gauss kernel function may overfit the data depending on the parameters, as is shown in Fig. 4.19. In function prediction using SVR/RBFN, it is sometimes difficult to determine appropriate parameters for kernel/basis function for given problems.

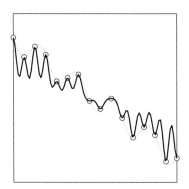

Fig. 4.19 Regression by μ-SVR

Combining a linear model and SVR/RBFN, it is expected that the burden of estimating the parameters in kernel/basis functions appropriately may be decreased. Using the same parameter with Fig. 4.19, Fig. 4.20 shows the result by the proposed model (4.9) with a linear polynomial model $g(\boldsymbol{x})$. Changing the parameters in μ-SVR, the root mean squared error (RMSE) is shown in Fig. 4.21. As can be readily seen from these figures, the method combining predetermined model and SVR/RBFN has a wider range of appropriate parameters, in other words we can select appropriate parameters more stably. Consequently, it is relatively easy to choose proper parameters in SVR/RBFN in the combined model. As a result, if we have some prior knowledge on the function to be predicted, we can decrease the burden of SVR/RBFN by utilizing the predetermined model based on the information.

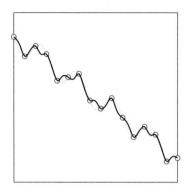

Fig. 4.20 Result by combining linear polynomial model and μ-SVR with the same parameter of Fig. 4.19

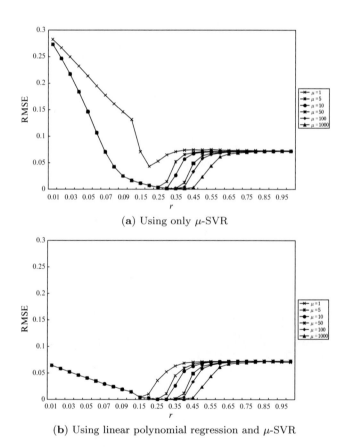

Fig. 4.21 Comparison of RMSE: r is a radius in Gauss kernel function

Chapter 5
Sequential Approximate Optimization

In many practical engineering design problems, the form of objective functions is not given explicitly in terms of design variables. Given the value of design variables, under this circumstance, the value of objective functions is obtained by some analysis such as structural analysis, fluid mechanic analysis, and thermodynamic analysis. Usually, these analyses are considerably time consuming to obtain a value of objective functions. In order to make the number of analyses as few as possible, *sequential approximate optimization* methods have been suggested [62, 71, 143, 155] (1) first, predicting the form of objective functions by techniques of machine learning (e.g., SVR, RBFN) and (2) optimizing the predicted objective function. A major problem in those methods is how to get a good approximation of the objective function based on as few sample data as possible. In addition, if the current solution is not satisfactory, it is needed to improve the approximation of the objective function by adding some additional data. The form of objective functions is revised by relearning on the basis of additional data step by step. Then, how to choose such additional data effectively becomes an important issue. In this section, we introduce several methods how to choose additional data, focusing on design of incremental experiments.

5.1 Metamodels

The identification of function forms of objective functions and constraints functions is referred to as "modeling" in practical problems. For the sake of simplicity, we consider throughout this section the following simple problem:

$$\text{minimize} \quad f(\boldsymbol{x}) \quad \text{over} \quad \boldsymbol{x} \in X \subset \mathbb{R}^n.$$

Then, the function f is also considered as a "model" of objectives. When it is difficult to identify the function form of f, but when we can observe the value of $f(\boldsymbol{x})$ for sampled point $\boldsymbol{x}$, we try to get an approximate function (or model) $\hat{f}$ for f on the basis of observations $(\boldsymbol{x}_i, y_i)$ where $y_i = f(\boldsymbol{x}_i)$ and $i = 1, \ldots, \ell$. This approximate function $\hat{f}$ is called a *metamodel* (in the sense of "model of the model"), a *surrogate model*, a *response surface*, a learning machine, etc., depending on research communities [12, 21, 70, 94, 95, 128, 155].

Now, our aim is to construct a good metamodel in the sense that:

1. We can obtain an approximate optimal solution through the metamodel with the property

$$|\hat{f}(\hat{\boldsymbol{x}}^*) - f(\boldsymbol{x}^*)| \leqq \epsilon_1,$$

where $\hat{\boldsymbol{x}}^*$ and $\boldsymbol{x}^*$ minimize $\hat{f}$ and f, respectively, and ϵ_1 is a given small positive number.

2. The total number of observations is as small as possible.

3. The metamodel $\hat{f}$ approximates well f entirely, if possible. Namely

$$\|\hat{f} - f\| \leqq \epsilon_2,$$

where ϵ_2 is a given small positive number.

If our aim is merely to find the optimal solution minimizing $f(\boldsymbol{x})$, then the metamodel $\hat{f}$ does not necessarily approximate well f entirely, but sufficiently well at least in a neighborhood of the optimal solution $\boldsymbol{x}^*$. Depending on practical problems, however, one may want to see the global behavior of the model f. Therefore, the priority of (3) above is behind the criteria (1) and (2) which are crucial in general.

In spite of the fact that the priority of (3) above is not so high to our aim in the book, the research from this viewpoint has a long history, e.g., in regression analysis, machine learning, etc. Therefore, we begin with our consideration on (3) and (2), and subsequently (1) as a specific feature along our purpose.

Note that we have typically two types of observations, namely

(a) With measurement error, e.g., in physical experiments

(b) Without any measurement error, e.g., in computational experiments in which one input causes only one output

Traditional statistics (in particular design of experiments, DoE, for considering (2)) and statistical learning theory (in particular active learning theory for considering (2)) treat the case of (a), while DACE (design analysis of computer experiments) [126, 128] does the case of (b). DASE (design analysis of simulation experiments) [70] with some artificial randomness in computer simulation is located intermediately between (a) and (b). We should change methods depending on the situation of our problems. However, the base of our discussion is put on the traditional statistic approach, because it is essential.

5.2 Optimal Design of Experiments

In the research field of design of experiments (DoE), there are numerous literatures on optimal design of experiments, e.g., [12,41,70,89,117,128]. In this section, we shall discuss how to select sample points (namely, experiments) for getting a better metamodel along this line.

Consider a traditional regression problem: find the best estimation of parameter $\boldsymbol{w}$ of metamodel $\hat{f}(\boldsymbol{x}, \boldsymbol{w})$ on the basis of observed data $(\boldsymbol{x}_i, y_i)$, $i = 1, \ldots, \ell$, with random error such that

$$y_i = \hat{f}(\boldsymbol{x}_i, \boldsymbol{w}) + \varepsilon_i, \quad E(\varepsilon_i) = 0, \quad \text{var}(\varepsilon_i) = \sigma, \quad \text{Cov}(\varepsilon_i \varepsilon_j) = 0. \tag{5.1}$$

Using the basis functions h_j, $j = 1, \ldots, m$, we assume

$$\hat{f}(\boldsymbol{x}, \boldsymbol{w}) = \sum_{j=1}^{m} w_j h_j(\boldsymbol{x}).$$

Letting $\boldsymbol{h} = (h_1, \ldots, h_m)^T$, $H = (\boldsymbol{h}(\boldsymbol{x}_1), \ldots, \boldsymbol{h}(\boldsymbol{x}_\ell))^T$, $\boldsymbol{y} = (y_1, \ldots, y_\ell)^T$ and $\boldsymbol{\varepsilon} = (\varepsilon_1, \ldots, \varepsilon_\ell)^T$, the regression model (5.1) can be rewritten in vector notations

$$\boldsymbol{y} = H\boldsymbol{w} + \boldsymbol{\varepsilon}, \qquad \boldsymbol{\varepsilon} \sim N(\boldsymbol{0}, \sigma^2 \mathbf{I}_\ell). \tag{5.2}$$

Then the squared error is given by

$$\mathcal{E}(\boldsymbol{w}) = ||\boldsymbol{y} - H\boldsymbol{w}||^2 = (\boldsymbol{y} - H\boldsymbol{w})^T (\boldsymbol{y} - H\boldsymbol{w}). \tag{5.3}$$

The solution $\hat{\boldsymbol{w}}$ minimizing (5.3) satisfies the following normal equation

$$H^T H \hat{\boldsymbol{w}} = H^T \boldsymbol{y}. \tag{5.4}$$

Therefore, assuming $H^T H$ to be nonsingular, the least square estimator $\hat{\boldsymbol{w}}$ is given by

$$\hat{\boldsymbol{w}} = (H^T H)^{-1} H^T \boldsymbol{y}.$$

In addition, the variance–covariance of $\hat{\boldsymbol{w}}$ is given by

$$\begin{aligned} \text{Cov}(\hat{\boldsymbol{w}}) &= \text{Cov}((H^T H)^{-1} H^T \boldsymbol{y}) \\ &= ((H^T H)^{-1} H^T) \text{Cov}(\boldsymbol{y})((H^T H)^{-1} H^T)^T \\ &= (H^T H)^{-1} H^T (\sigma^2 \mathbf{I}) H (H^T H)^{-1} \\ &= (H^T H)^{-1} H^T H (H^T H)^{-1} \sigma^2 \\ &= (H^T H)^{-1} \sigma^2, \end{aligned}$$

and an unbiased estimate of σ^2 is taken as

$$\hat{\sigma}^2 = \frac{1}{\ell - m - 1}(\boldsymbol{y} - H\hat{\boldsymbol{w}})^T(\boldsymbol{y} - H\hat{\boldsymbol{w}}).$$

The linear least square estimator $\hat{\boldsymbol{w}}$ has an interesting property. If the outputs $\boldsymbol{y}$ are normally distributed, then this estimator $\hat{\boldsymbol{w}}$ is also normally distributed. Combining this property of mean and covariance gives

$$\boldsymbol{w} \sim N(\hat{\boldsymbol{w}}, (H^T H)^{-1}\sigma^2).$$

The predicted response at a point $\boldsymbol{x}_0$ is $\hat{y}(\boldsymbol{x}_0) = \boldsymbol{h}^T(\boldsymbol{x}_0)\hat{\boldsymbol{w}}$, where $\boldsymbol{h}^T(\boldsymbol{x}_0) := (h_1(\boldsymbol{x}_0), \ldots, h_m(\boldsymbol{x}_0))$. Then the *prediction variance* is given by

$$v(\boldsymbol{x}_0) := \mathrm{Var}(\hat{y}(\boldsymbol{x}_0)) = \boldsymbol{h}^T(\boldsymbol{x}_0)(H^T H)^{-1}\boldsymbol{h}(\boldsymbol{x}_0)\sigma^2. \tag{5.5}$$

Remark 5.1. As was noted in Sect. 4.1, if $H^T H$ is singular, we impose some regularization on it, e.g., for sufficiently small $\lambda > 0$, we use $H^T H + \lambda I$ instead of $H^T H$.

Define an order for matrices A and B induced by some positive semidefinite matrix P in such a way that

$$A \leq B \text{ if } B = A + P. \tag{5.6}$$

Then, it is known (Gauss–Markov theorem [89, 121, 137]) that for any design the least square estimator $\hat{\boldsymbol{w}}$ is the best one in the sense that

$$\mathrm{Cov}(\hat{\boldsymbol{w}}) = \min_{\tilde{\boldsymbol{w}}} \mathrm{Cov}(\tilde{\boldsymbol{w}}),$$

where $\tilde{\boldsymbol{w}}$ denotes any linear unbiased estimators and the order of the variance–covariance matrices Cov follows (5.6).

Now, we try to minimize $\mathrm{Cov}(\hat{\boldsymbol{w}})$ by selecting the most appropriate design among all possible designs $\boldsymbol{\xi}_\ell := \{\boldsymbol{x}_1, \ldots, \boldsymbol{x}_\ell\}$. Here we denote the *information matrix* $H^T H$ by $M(\boldsymbol{\xi}_\ell)$ because it depends on the design $\boldsymbol{\xi}_\ell := \{\boldsymbol{x}_1, \ldots, \boldsymbol{x}_\ell\}$.[1] Since $\mathrm{Cov}(\hat{\boldsymbol{w}}) = \sigma^2 M(\boldsymbol{\xi}_\ell)^{-1}$, our aim is formulated as

$$(P_1) \qquad \text{minimize} \quad M(\boldsymbol{\xi}_\ell)^{-1} \quad \text{over} \quad \boldsymbol{\xi}_\ell = \{\boldsymbol{x}_1, \ldots, \boldsymbol{x}_\ell\}.$$

This is a kind of vector optimization with respect to the order for matrices defined by (5.6). Like usual multiobjective optimization, we have several extremal solutions for (P_1) depending on scalarization. The most popular way of "scalarization" for (P_1) is to consider the confidence region.

---

[1] In a formal sense, the number of experiments is also a decision variable. However, since it is mostly given beforehand in many case, we may consider cases with a given number of experiments ℓ.

5.2 Optimal Design of Experiments   117

Rewriting the squared error given by (5.3), we have

$$\mathcal{E}(\boldsymbol{w}) = (\boldsymbol{y} - H\hat{\boldsymbol{w}})^T (\boldsymbol{y} - H\hat{\boldsymbol{w}}) + \sigma^2 (\boldsymbol{w} - \hat{\boldsymbol{w}})^T M(\boldsymbol{\xi}_\ell)(\boldsymbol{w} - \hat{\boldsymbol{w}}).$$

Consider optimality criteria in the parameter space. Note that

$$(\boldsymbol{w} - \hat{\boldsymbol{w}})^T M(\boldsymbol{\xi}_\ell)(\boldsymbol{w} - \hat{\boldsymbol{w}}) \leqq \sigma^2 \chi_\alpha^2(k),$$

where $\chi_\alpha^2(k)$ is the chi-square α quantile with k degrees of freedom, defines a *confidence ellipsoid* [48, 123]. Therefore, our aim is to minimize the size of ellipsoid by selecting the most appropriate design $\boldsymbol{\xi}_\ell$. A most intuitive measure for the size of ellipsoid is its volume which is proportional to $|M|^{-1/2}$ (see, e.g., [136, p. 679]) where $|A|$ denotes the determinant of matrix A. This leads to *D-optimality* criterion named after "determinant."

Another one is given by the length of the largest principal axis of the ellipsoid. In general, we have for any non-zero θ of $\mathbb{R}^m$,

$$\lambda_{\min}(M(\boldsymbol{\xi}_\ell)) \leqq \frac{\theta^T M(\boldsymbol{\xi}_\ell)\theta}{\theta^T \theta} \leqq \lambda_{\max}(M(\boldsymbol{\xi}_\ell)). \tag{5.7}$$

where $\lambda_{\min}(M(\boldsymbol{\xi}_\ell))$ and $\lambda_{\max}(M(\boldsymbol{\xi}_\ell))$ denote the minimal eigenvalue of $M(\boldsymbol{\xi}_\ell)$ and the maximal eigenvalue of $M(\boldsymbol{\xi}_\ell)$, respectively. The length of the largest principal axis of the ellipsoid is proportional to $\lambda_{\min}^{-1/2}(M(\boldsymbol{\xi}_\ell))$ and that of the smallest one is to $\lambda_{\max}^{-1/2}(M(\boldsymbol{\xi}_\ell))$. Therefore, the criterion corresponding to the length of the largest principal axis of the ellipsoid is given by $\lambda_{\min}^{-1/2}$. This is the so-called *E-criterion*. The name of E-optimality criterion is originated from "extreme" in the above sense.

Unifying the notation by using M^{-1}, we have by virtue of the fact that the eigenvalue of M^{-1} is given by $1/\lambda$ for the eigenvalue λ of M:

1. *D-optimality*: minimize $|M^{-1}|$

2. *E-optimality*: minimize $\lambda_{\max}(M^{-1})$

In addition, a criterion may be given by the length of the diagonal of the minimal rectangle including the ellipsoid which corresponds to $\sum_{i=1}^m 1/\lambda_i = \text{tr}(M^{-1})$, where λ_i, $i = 1, \ldots, m$ are eigenvalues of M and the operator "tr" denotes the trace which is defined as the sum of diagonal elements of the matrix. This criterion is called *A-optimality* criterion in which "A" stands for "average." Therefore,

3. *A-optimality*: minimize $\text{tr}(M^{-1})$

The above criterion (3) means to minimize the average of diagonal elements of M^{-1}, namely the average of individual variances of the estimated parameters (Fig. 5.1).

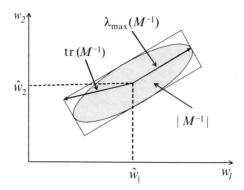

Fig. 5.1 Confidence ellipsoid and several criteria for optimal design [48]

With the matrix M, we have the following equivalent criteria:

1'. *D-optimality*: maximize $|M|$
2'. *E-optimality*: maximize $\lambda_{\min}(M)$
3'. *A-optimality*: maximize $(\mathrm{tr}(M^{-1}))^{-1}$

Remark 5.2. {D-, E-, A-} optimality criteria are all based on the variance of estimation error. On the other hand, there are other criteria based on the variance of prediction error. Since the prediction variance at $\bm{x}_0$ is given by (5.5), one may easily consider the following:

4. G-optimality: minimize $\max\limits_{\bm{x}_0 \in X} v(\bm{x}_0)$
5. V-optimality: minimize $\operatorname*{ave}\limits_{\bm{x}_0 \in X} v(\bm{x}_0)$

where X is a set of all possible $\bm{x}_0$ and *ave* denotes the average. Note that it is not so easy to compute {G-, V-} optimality criteria.

Remark 5.3. Although there have been developed many other criteria, D-optimality criterion is most popularly used in practice. An advantage of D-optimality criterion is that it does not depend on the scale of the variables, because linear transformations leave the D-optimal design unchanged, while not necessarily the case for A- and E-optimal designs. For designs with all factors qualitative, however, the problem of scale does not arise and hence A- and E-optimal designs are frequently employed.

Example 5.1. Consider the metamodel using linear polynomial basis functions

$$\hat{f}_1(x, \bm{w}) = w_0 + w_1 x.$$

For the sake of simplicity, suppose that $-1 \leqq x \leqq 1$. Consider designs $\bm{\xi}$ with five observation points x_1, x_2, x_3, x_4, x_5, and three designs

5.2 Optimal Design of Experiments

$$\xi_1 = \{x_1 = -1, \ x_2 = -1, \ x_3 = -1, \ x_4 = 1, \ x_5 = 1\},$$

$$\xi_2 = \left\{x_1 = -1, \ x_2 = -\frac{1}{2}, \ x_3 = 0, \ x_4 = \frac{1}{2}, \ x_5 = 1\right\},$$

$$\xi_3 = \{x_1 = -1, \ x_2 = -1, \ x_3 = 0, \ x_4 = 1, \ x_5 = 1\}.$$

Then, what is the best among them with regard to {D-,E-,A-} optimality criterion? The information matrix M is given by

$$M = \begin{pmatrix} 1 & 1 & 1 & 1 & 1 \\ x_1 & x_2 & x_3 & x_4 & x_5 \end{pmatrix} \begin{pmatrix} 1 & x_1 \\ 1 & x_2 \\ 1 & x_3 \\ 1 & x_4 \\ 1 & x_5 \end{pmatrix},$$

so that

$$M = \begin{pmatrix} 5 & x_1 + x_2 + x_3 + x_4 + x_5 \\ x_1 + x_2 + x_3 + x_4 + x_5 & x_1^2 + x_2^2 + x_3^2 + x_4^2 + x_5^2 \end{pmatrix}.$$

The determinants of M for three designs ξ_1, ξ_2 and ξ_3 are given by

$$|M(\xi_1)| = 24.0, \quad |M(\xi_2)| = 12.5, \quad |M(\xi_3)| = 20.0.$$

Regarding the minimal eigenvalue, we have

$$\lambda_{min} M(\xi_1) = 4.0, \quad \lambda_{min} M(\xi_2) = 2.5, \quad \lambda_{min} M(\xi_3) = 4.0.$$

Furthermore, the traces of M are

$$(\operatorname{tr} M^{-1}(\xi_1))^{-1} = 2.40, \ (\operatorname{tr} M^{-1}(\xi_2))^{-1} = 1.67, \ (\operatorname{tr} M^{-1}(\xi_3))^{-1} = 2.22.$$

Therefore, we have the same order among the design ξ_1, ξ_2 and ξ_3 with regard to any of D-, E-, A-optimality criteria.

Example 5.2. In case of metamodel with quadratic polynomial basis functions given by

$$\hat{f}_2(x, \boldsymbol{w}) = w_0 + w_1 x + w_2 x^2,$$

we have

$$H = \begin{pmatrix} 1 & x_1 & x_1^2 \\ 1 & x_2 & x_2^2 \\ 1 & x_3 & x_3^2 \\ 1 & x_4 & x_4^2 \\ 1 & x_5 & x_5^2 \end{pmatrix}.$$

Since $M = H^T H$, the value of each optimality criterion for the same three designs $\boldsymbol{\xi}_1$, $\boldsymbol{\xi}_2$ and $\boldsymbol{\xi}_3$ as in Example 5.1 is given by

$$
\begin{array}{lll}
|M(\boldsymbol{\xi}_1)| = 0, & |M(\boldsymbol{\xi}_2)| = 10.9375, & |M(\boldsymbol{\xi}_3)| = 16, \\
\lambda_{min} M(\boldsymbol{\xi}_1) = 0.0, & \lambda_{min} M(\boldsymbol{\xi}_2) = 0.6787, & \lambda_{min} M(\boldsymbol{\xi}_3) = 0.4689, \\
(\text{tr } M^{-1}(\boldsymbol{\xi}_1))^{-1} = 0.000, & (\text{tr } M^{-1}(\boldsymbol{\xi}_2))^{-1} = 0.493, & (\text{tr } M^{-1}(\boldsymbol{\xi}_3))^{-1} = 0.4.
\end{array}
$$

Therefore, which design is the best for the quadratic polynomial model depends on the optimality criteria.

Example 5.3. For the metamodel using Gaussian basis functions

$$
h_j(x) = \exp\left(-\frac{(x - x_j)^2}{2r^2}\right),
$$

where x_j is the observation point of each experiment $\boldsymbol{\xi}$, we have

$$
\begin{array}{lll}
|M(\boldsymbol{\xi}_1)| = 0, & |M(\boldsymbol{\xi}_2)| = 0.007, & |M(\boldsymbol{\xi}_3)| = 0, \\
\lambda_{min} M(\boldsymbol{\xi}_1) = 0, & \lambda_{min} M(\boldsymbol{\xi}_2) = 0.0087, & \lambda_{min} M(\boldsymbol{\xi}_3) = 0, \\
(\text{tr } M^{-1}(\boldsymbol{\xi}_1))^{-1} = 0, & (\text{tr } M^{-1}(\boldsymbol{\xi}_2))^{-1} = 0.0079, & (\text{tr } M^{-1}(\boldsymbol{\xi}_3))^{-1} = 0.
\end{array}
$$

Therefore, the {D-, E-, A-} optimality criteria for the Gaussian model yields the optimal design for which sample points are distributed almost evenly.

5.3 Distance-Based Criteria for Optimal Design

Optimal designs based on the so-called *alphabetical optimality* criteria stated in the previous section depend on the underlying model. It is intuitively seen that the more widely the observed data are spread, the better approximate the metamodel is. From this idea, *space filling methods* are developed. The simplest example may be the *mesh-grid method*. In view of decreasing the number of experiments, *Latin hypercube* (LH) method is more widely applied. However, our main concern is incremental design in which we start with a relatively small number of experiments and then add experiments in sequence, if necessary. To this aim, mesh-grid method and LH method are not appropriate. Since the *distance-based criteria* proposed by Johnson–Moore–Ylvisaker [63] can be applied to incremental design, we will discuss it in more detail in the following.

In general, the distance between a point $\boldsymbol{x}$ and a set S is defined as

$$
d(\boldsymbol{x}, S) := \min_{\boldsymbol{s} \in S} ||\boldsymbol{x} - \boldsymbol{s}||.
$$

5.3 Distance-Based Criteria for Optimal Design

Let $\mathcal{D}$ be a set of all possible designs, which is closed and bounded in the design space (usually $\mathbb{R}^n$). In Sect. 5.2, a design with ℓ experiment points was denoted by $\boldsymbol{\xi} = \{\boldsymbol{x}_1, \ldots, \boldsymbol{x}_\ell\}$. Since designs are subsets of $\mathcal{D}$, we denote candidate designs with ℓ design points in $\mathcal{D}$ by $\mathcal{C}$ here. Then we have:

1. *Max–min distance optimality*: maximize $\min_{\boldsymbol{x} \in \mathcal{C}} d(\boldsymbol{x}, \mathcal{C} \backslash \boldsymbol{x})$ with respect to $\mathcal{C}$
2. *Min–max distance optimality*: minimize $\max_{\boldsymbol{x} \in \mathcal{D}} d(\boldsymbol{x}, \mathcal{C})$ with respect to $\mathcal{C}$

Max–min distance designs tend to cover the design space as much as possible, because no two points should be too close to each other. The distance $\min_{\boldsymbol{x} \in \mathcal{C}} d(\boldsymbol{x}, \mathcal{C} \backslash \boldsymbol{x})$ can be interpreted to indicate the degree of covering by the candidate design $\mathcal{C}$. On the other hand, min–max distance designs tend to spread out in the design space as uniformly as possible, because we can interpret that points outside of $\mathcal{C}$ pull out $\mathcal{C}$ as much as possible.

These distance-based criteria yield a design which is distributed uniformly and fill the design space as much as possible. Moreover, note that they are not dependent on the underlying model.

Example 5.4. Let $\mathcal{D}$ be a rectangle of $[0,1] \times [0,1]$. Consider designs with seven sample points as given in [63]. Figure 5.2a shows the max–min optimal design using Euclidean distance, while Fig. 5.2b the min–max optimal design using Euclidean distance. As is readily seen, the max–min distance design tend to place sample points near the boundary of possible design set $\mathcal{D}$.

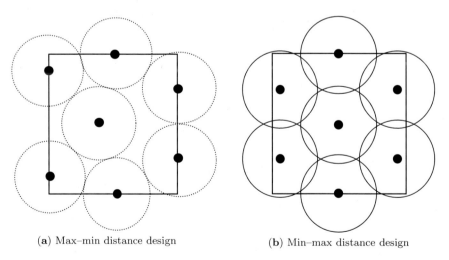

(a) Max–min distance design   (b) Min–max distance design

Fig. 5.2 Max–min and min–max distance criteria for optimal design [63]

5.4 Incremental Design of Experiments

In sequential approximate optimization, it is desirable to obtain an approximate optimal solution by as less number of experiments as possible. To this aim, we usually start with a relatively small number of experiments and add experiments sequentially, if necessary. In the following, we discuss the above several optimality criteria from a viewpoint of this *incremental design*.

Suppose that we have made experiments at $x_1, \ldots, x_\ell$ with their observations $y_1, \ldots, y_\ell$, respectively, so far. In addition, suppose that we need to make a new experiment $x_{\ell+1}$ so that our aim may be attained better. It will be shown in the following how the selected $x_{\ell+1}$ differs depending on the criterion.

Let our metamodel be $\hat{f}(x, w) = \sum_{j=1}^{m} w_j h_j(x)$. Then the design matrix with experiments $x_1, \ldots, x_\ell, x_{\ell+1}$ is given by

$$H_{\ell+1} = \begin{pmatrix} h_1(x_1) & h_2(x_1) & \cdots & h_m(x_1) \\ h_1(x_2) & h_2(x_2) & \cdots & h_m(x_2) \\ \vdots & \vdots & \ddots & \vdots \\ h_1(x_\ell) & h_2(x_\ell) & \cdots & h_m(x_\ell) \\ h_1(x_{\ell+1}) & h_2(x_{\ell+1}) & \cdots & h_m(x_\ell + 1) \end{pmatrix}.$$

In order to find the most appropriate additional experiment $x_{\ell+1}$, we solve optimization with some optimality criterion. For example, applying D-optimality criteria, our aim is to find $x_{\ell+1}$ minimizing $|M_{\ell+1}^{-1}|$ over the design set $\mathcal{D}$.

Remark 5.4. Usually it is difficult to find an optimal $x_{\ell+1}$ over the whole design set $\mathcal{D}$ in terms of above criteria. Then a practical method for finding an approximate optimal $x_{\ell+1}$ is to select the best one among several randomly generated candidates.

Remark 5.5. As the incrementation proceeds, the size of matrix to be inversed becomes large, which causes the matrix inversion to be time consuming.

The following update formula is convenient for sequential matrix inversion: Let $h_j^T = (h_1(x_j), \ldots, h_m(x_j))$ and $H_\ell = (h_1, \ldots, h_\ell)^T$. Adding a new experiment $x_{\ell+1}$, we have

$$H_{\ell+1} = \begin{bmatrix} H_\ell \\ h_{\ell+1}^T \end{bmatrix}.$$

For $M_\ell = H_\ell^T H_\ell$ and $M_{\ell+1} = H_{\ell+1}^T H_{\ell+1}$,

$$M_{\ell+1}^{-1} = M_\ell^{-1} - \frac{M_\ell^{-1} h_{\ell+1} h_{\ell+1}^T M_\ell^{-1}}{1 + h_{\ell+1}^T M_\ell^{-1} h_{\ell+1}}.$$

Several sequential update formulae for determinant and trace of matrix have been developed (see, e.g., [41]).

5.4 Incremental Design of Experiments

When using a distance-based criteria, a simple method for incremental design using k-NN method has been proposed [106]: In Fig. 5.3, the circle points are existing experiments. Three triangles are candidates for new experiments. Calculate, for example, the shortest distance between each candidate experiment and the set of existing experiments. We select the candidate with the maximal value of this shortest distance to existing experiments as the one which is in less populated area. In Fig. 5.3, the white triangle is judged as the best one among the candidates. This idea can be easily extended to a method taking into account k-shortest distances [106].

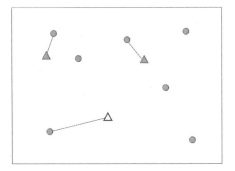

Fig. 5.3 Additional point by distance-based optimality criteria

We will compare the results for three metamodels for $x \in \mathbb{R}^2$ using the following basis functions:

1. Linear polynomial

$$h_1(x) = 1, \; h_2(x) = x_1, \; h_3(x) = x_2$$

2. Quadratic polynomial

$$h_1(x) = 1, \; h_2(x) = x_1, \; h_3(x) = x_2,$$
$$h_4(x) = x_1^2, \; h_5(x) = x_2^2, \; h_6(x) = x_1 x_2$$

3. Gaussian

$$h_j(x) = \exp\left(-\frac{||x - x_j||^2}{2r^2}\right), \; j = 1, \ldots, \ell$$

In Figs. 5.4–5.9, we show the results by the D-optimality criterion. We start with initial four experiments at four corners. We add experiments one by one on the basis of D-optimality criteria. In this event, 100 candidates are randomly generated for each selection of additional experiment, and are evaluated by the determinant of information matrix. It is seen from Fig. 5.9 that additional experiments for models using linear and quadratic polynomial basis functions are biased in some particular area, while the ones for Gaussian basis functions are almost uniformly distributed. Namely, a new experiment based on D-optimality for the metamodel using Gaussian basis functions is allocated in less populated area at each selection.

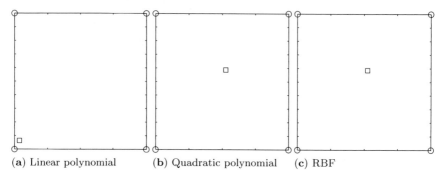

(a) Linear polynomial   (b) Quadratic polynomial   (c) RBF

Fig. 5.4 D-optimal designs for several models (no. of sample data = 4)

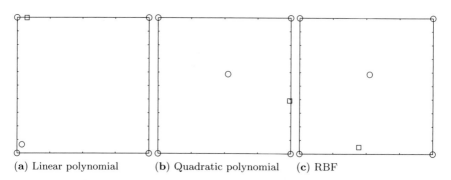

(a) Linear polynomial   (b) Quadratic polynomial   (c) RBF

Fig. 5.5 D-optimal designs for several models (no. of sample data = 5)

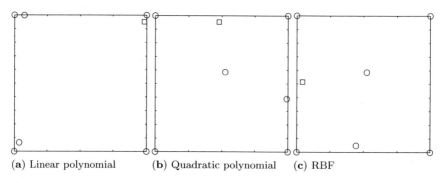

(a) Linear polynomial   (b) Quadratic polynomial   (c) RBF

Fig. 5.6 D-optimal designs for several models (no. of sample data = 6)

5.4 Incremental Design of Experiments

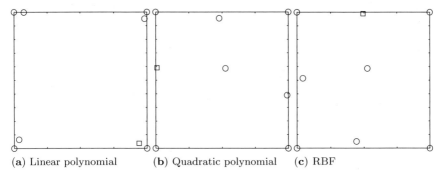

Fig. 5.7 D-optimal designs for several models (no. of sample data = 7)

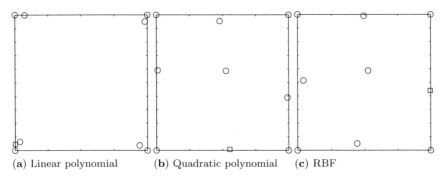

Fig. 5.8 D-optimal designs for several models (no. of sample data = 8)

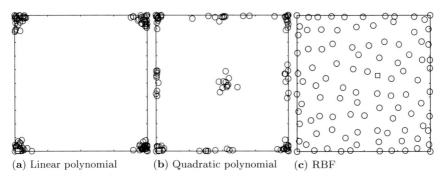

Fig. 5.9 D-optimal designs for several models (no. of sample data = 100)

Employing the distance-based criteria for incremental design stated above, we have results shown in Fig. 5.10. One may see that additional sample points are selected in less populated area at each incrementation than D-optimality with Gaussian model. This means the distance-based criterion performs effectively aiming at space filling.

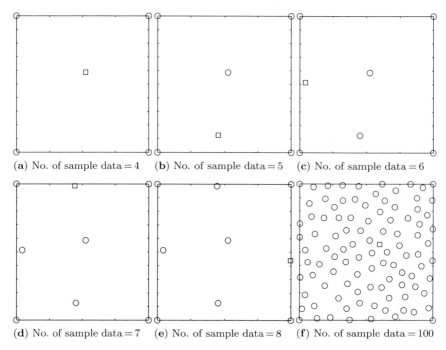

Fig. 5.10 Distance-based designs for incremental experiments

Remark 5.6. Depending on metamodels, the additional sample points differ. There are some cases in which we can utilize linear/quadratic models sometimes on the basis of a priori knowledge. In such a particular case, it is reasonable that the sample data are allocated in some particular area, because they can provide "good" metamodel with regard to variance. However, in cases where we cannot utilize a priori knowledge on models, and also in highly nonlinear cases, Gaussian basis functions can be used to approximate models well, as already seen in RBF networks. It seems reasonable that a good approximate model for highly nonlinear models can be obtained by space filling sample points. Also one may see that the distance-based criterion, which does not depend on models, tends to cover the design space better than other criteria.

5.5 Kriging and Efficient Global Optimization

The *Kriging* is a method for an optimal spatial prediction, and there are several kinds of Kriging such as simple Kriging, ordinary Kriging, universal Kriging, and so on. The Kriging was originally developed in geostatistics (also known as spatial statistics) by the South African mining engineer Krige in his

5.5 Kriging and Efficient Global Optimization   127

master's thesis [76] on analyzing mining data in the 1950s. The Kriging has been popularized as one of stochastic methods for spatial prediction. Besides, theoretical results were developed by Matheron [87] and for more extensive reviews, the readers can be referred to $[26, 66, 86, 124]$.

Later, the Kriging, as a new method of model prediction $[126, 156]$, has been often used for sequential approximate optimization with design and analysis of computer experiments (DACE) in which no measurement error occurs $[64, 65, 134, 135]$.

5.5.1 Kriging and Prediction

In order to introduce Kriging approach in DACE, consider the following stochastic process model:

$$Y(\boldsymbol{x}) = \sum_{j=1}^{m} w_j h_j(\boldsymbol{x}) + Z(\boldsymbol{x}), \qquad (5.8)$$

where $h_j(\boldsymbol{x})$, $j = 1, \ldots, m$, are deterministic basis functions, w_j are corresponding coefficients, and $Z(\boldsymbol{x})$ is a *Gaussian process*.

Especially, $E(Z(\boldsymbol{x})) = 0$ and the covariance between the Zs at two points $\boldsymbol{x} = (x_1, \ldots, x_n)^T$ and $\boldsymbol{x}' = (x'_1, \ldots, x'_n)^T$ is given by

$$\mathrm{Cov}(Z(\boldsymbol{x}), Z(\boldsymbol{x}')) = \sigma_Z^2 R(\boldsymbol{x}, \boldsymbol{x}'),$$

where σ_Z^2 is the *process variance* with a scalar value and $R(\cdot, \cdot)$ is a *correlation function*. Note that w_j, $j = 1, \ldots, m$ and σ_Z^2 can be derived from given sample data, and their estimation will be described later.

For computational convenience, a correlation function is supposed to follow the *product correlation rule*

$$R(\boldsymbol{x}, \boldsymbol{x}') = \prod_{i=1}^{n} R_i(x_i - x'_i).$$

There are several types of correlation functions such as linear correlation function, exponential correlation function, Gaussian correlation function, and so on. Among them, the following correlation function family is commonly used:

$$R(\boldsymbol{x}, \boldsymbol{x}') = \prod_{i=1}^{n} \exp\left(-\frac{|x_i - x'_i|^{p_i}}{2 r_i^2}\right),$$

where $r_i > 0$ and $0 < p_i \leqq 2$ are the parameters. The p_is denote the smoothness of the correlation parameters, while smaller r_i indicates nonlinearity. We use $p_i = 2$ in the simulations of this book.

Remark 5.7. As stated above, it is seen that the covariance between the Zs at any two points $\boldsymbol{x}$ and $\boldsymbol{x}'$ depends only on $\|\boldsymbol{x} - \boldsymbol{x}'\|$, i.e., on their relative location. In many statistical linear regression model, $Z(\boldsymbol{x})$ is assumed to be independent error term. However, computer experiments are distinct from physical experiments in view of the fact that they produce the same value of output for the same value of input. Since there is no measurement error or noise in DACE, assuming that $Z(\boldsymbol{x})$ is an independent error term is not suitable in modeling of DACE. It is reasonable to regard $Z(\boldsymbol{x})$ as modeling error term, which is the essential idea of Kriging, and to assume that $Z(\boldsymbol{x})$ and $Z(\boldsymbol{x}')$ are correlated according to the relative location of two points $\boldsymbol{x}$ and $\boldsymbol{x}'$, namely not independent. In other words, if the distance between $\boldsymbol{x}$ and $\boldsymbol{x}'$ is close, the relation between $Z(\boldsymbol{x})$ and $Z(\boldsymbol{x}')$ is close.

Model Prediction

An estimation method using the *maximum likelihood* is the most popular technique in statistics. The *maximum likelihood estimators* (MLE) have several theoretical properties, such as invariance, sufficiency and asymptotic properties, and so on. Especially, the MLE, if this exists, is equivalent to the *best linear unbiased estimation* under the assumption of Gaussian distribution. Using the MLE, we explain to derive unknown parameters $\boldsymbol{w} = (w_1, \ldots, w_m)^T$ and σ_Z^2 in the model (5.8).

For simplicity, using matrix notations for given sample data $(\boldsymbol{x}_i, y_i)$, $i = 1, \ldots, \ell$,

$$\boldsymbol{y} = \begin{pmatrix} y_1 \\ \vdots \\ y_\ell \end{pmatrix}, \quad H = \begin{pmatrix} h_1(\boldsymbol{x}_1) \ldots h_m(\boldsymbol{x}_1) \\ \vdots \quad \ddots \quad \vdots \\ h_1(\boldsymbol{x}_\ell) \ldots h_m(\boldsymbol{x}_\ell) \end{pmatrix},$$

$$\boldsymbol{w} = \begin{pmatrix} w_1 \\ \vdots \\ w_m \end{pmatrix}, \quad \mathbf{z} = \begin{pmatrix} Z(\boldsymbol{x}_1) \\ \vdots \\ Z(\boldsymbol{x}_\ell) \end{pmatrix},$$

we have the following equation from the model (5.8):

$$\boldsymbol{y} = H\boldsymbol{w} + \mathbf{z}. \tag{5.9}$$

A correlation matrix $\mathbf{R}$ for given sample points $\boldsymbol{x}_i$, $i = 1, \ldots, \ell$, is denoted by

$$\mathbf{R} = \begin{pmatrix} R(\boldsymbol{x}_1, \boldsymbol{x}_1) \ldots R(\boldsymbol{x}_1, \boldsymbol{x}_\ell) \\ \vdots \quad \ddots \quad \vdots \\ R(\boldsymbol{x}_\ell, \boldsymbol{x}_1) \ldots R(\boldsymbol{x}_\ell, \boldsymbol{x}_\ell), \end{pmatrix},$$

5.5 Kriging and Efficient Global Optimization

and for untried point $\boldsymbol{x}$, the following notations are used:

$$\mathbf{h_x} = \begin{pmatrix} h_1(\boldsymbol{x}) \\ \vdots \\ h_m(\boldsymbol{x}) \end{pmatrix}, \quad \mathbf{r_x} = \begin{pmatrix} R(\boldsymbol{x}_1, \boldsymbol{x}) \\ \vdots \\ R(\boldsymbol{x}_\ell, \boldsymbol{x}) \end{pmatrix},$$

where $\mathbf{r_x}$ is a vector of correlations between the Zs at $\boldsymbol{x}$ and $\boldsymbol{x}_1, \ldots, \boldsymbol{x}_\ell$.

In order to find the maximum likelihood estimates of the unknown parameters $\boldsymbol{w}$ and σ_Z^2 for given values of θ_i and r_i, $i = 1, \ldots, n$, consider the log-likelihood up to an additive constant

$$\ln L(\boldsymbol{w}, \sigma_Z^2) = -\frac{1}{2}\left(\ell \ln \sigma_Z^2 + \ln |\mathbf{R}| + \frac{(\boldsymbol{y} - H\boldsymbol{w})^T \mathbf{R}^{-1}(\boldsymbol{y} - H\boldsymbol{w})}{\sigma_Z^2}\right). \quad (5.10)$$

Differentiating (5.10) with respect to $\boldsymbol{w}$ and σ_Z^2 and setting that the results equal to zero, the following holds:

$$\frac{\partial \ln L(\boldsymbol{w}, \sigma_Z^2)}{\partial \boldsymbol{w}} = \frac{1}{\sigma_Z^2}\left(H^T \mathbf{R}^{-1} \boldsymbol{y} - H^T \mathbf{R}^{-1} H \boldsymbol{w}\right) = \mathbf{0},$$

$$\frac{\partial \ln L(\boldsymbol{w}, \sigma_Z^2)}{\partial \sigma_Z^2} = -\frac{\ell}{2\sigma_Z^2} + \frac{1}{2\sigma_Z^4}(\boldsymbol{y} - H\boldsymbol{w})^T \mathbf{R}^{-1}(\boldsymbol{y} - H\boldsymbol{w}) = 0,$$

which yield

$$H^T \mathbf{R}^{-1} \boldsymbol{y} - H^T \mathbf{R}^{-1} H \boldsymbol{w} = \mathbf{0},$$

$$\sigma_Z^2 - \frac{1}{\ell}(\boldsymbol{y} - H\boldsymbol{w})^T \mathbf{R}^{-1}(\boldsymbol{y} - H\boldsymbol{w}) = 0.$$

Therefore, the MLE of $\boldsymbol{w}$ is the *generalized least squares estimator*[2]

$$\hat{\boldsymbol{w}} = (H^T \mathbf{R}^{-1} H)^{-1} H^T \mathbf{R}^{-1} \boldsymbol{y}, \quad (5.11)$$

and the MLE of σ_Z^2

$$\hat{\sigma}_Z^2 = \frac{1}{\ell}(\boldsymbol{y} - H\hat{\boldsymbol{w}})^T \mathbf{R}^{-1}(\boldsymbol{y} - H\hat{\boldsymbol{w}}). \quad (5.12)$$

Mean and Variance of Prediction

Next, the mean and error of prediction in the model (5.8) are estimated by using the best linear unbiased predictor. To begin with, we introduce the *best linear unbiased predictor* commonly used in statistics.

---

[2] A method for estimating parameters $w_1, \ldots, w_m$ linearly appearing when the prediction errors are not independent but their variance–covariance matrix is known up to a constant [1].

Consider a predictor $y(\boldsymbol{x})$ of $Y(\boldsymbol{x})$ given by a *linear predictor*

$$y(\boldsymbol{x}) = \sum_{i=1}^{\ell} c_i(\boldsymbol{x}) y_i := \mathbf{c}_{\boldsymbol{x}}^T \boldsymbol{y} \tag{5.13}$$

with the *unbiasedness* $E(y(\boldsymbol{x})) = E(Y(\boldsymbol{x}))$. Then, a predictor $y(\boldsymbol{x})$ becomes a *linear unbiased predictor*, and we can define the best linear unbiased predictor (BLUP) as follows:

Definition 5.1 (Best Linear Unbiased Predictor). A predictor $y(\boldsymbol{x})$ is called the *best linear unbiased predictor* (BLUP) if the mean squared error $E(y(\boldsymbol{x}) - Y(\boldsymbol{x}))^2$ is minimized among all linear unbiased predictors.

In order to determine the BLUP of $Y(\boldsymbol{x})$ described in the model (5.8), consider a measure of the prediction error to quantify the covariance and a constraint to ensure the unbiasedness.

Using the definition of linear predictor (5.13), from (5.9) and the assumption of $E(Z(\boldsymbol{x})) = 0$,

$$E(y(\boldsymbol{x})) = E\left(\mathbf{c}_{\boldsymbol{x}}^T \boldsymbol{y}\right) = \mathbf{c}_{\boldsymbol{x}}^T H \boldsymbol{w} \tag{5.14}$$

for any coefficient $\mathbf{c}_{\boldsymbol{x}}$.

In a similar way, the following holds:

$$E(Y(\boldsymbol{x})) = E\left(\sum_{j=1}^{m} w_j h_j(\boldsymbol{x})\right) + E(Z(\boldsymbol{x})) = \mathbf{h}_x^T \boldsymbol{w} \tag{5.15}$$

by the assumptions of the model (5.8) that h_j, $j = 1, \ldots, m$ are deterministic and $E(Z(\boldsymbol{x})) = 0$.

From above two formulations (5.14) and (5.15), we obtain the unbiasedness constraint

$$H^T \mathbf{c}_{\boldsymbol{x}} = \mathbf{h}_{\boldsymbol{x}}. \tag{5.16}$$

For a linear predictor $y(\boldsymbol{x}) = \mathbf{c}_{\boldsymbol{x}}^T \boldsymbol{y}$ of $Y(\boldsymbol{x})$, the mean squared error (MSE) of prediction is given by

$$\begin{aligned}
E\left(\mathbf{c}_{\boldsymbol{x}}^T \boldsymbol{y} - Y(\boldsymbol{x})\right)^2 &= E\left(\mathbf{c}_{\boldsymbol{x}}^T \boldsymbol{y} \boldsymbol{y}^T \mathbf{c}_{\boldsymbol{x}} + Y^2(\boldsymbol{x}) - 2\mathbf{c}_{\boldsymbol{x}}^T \boldsymbol{y} Y(\boldsymbol{x})\right) \\
&= E\left(\mathbf{c}_{\boldsymbol{x}}^T (H\boldsymbol{w} + \mathbf{z})(H\boldsymbol{w} + \mathbf{z})^T \mathbf{c}_{\boldsymbol{x}} + (\mathbf{h}_x^T \boldsymbol{w} + Z(\boldsymbol{x}))^2 \right. \\
&\qquad \left. -2\mathbf{c}_{\boldsymbol{x}}^T (H\boldsymbol{w} + \mathbf{z})(\mathbf{h}_x^T \boldsymbol{w} + Z(\boldsymbol{x}))\right) \\
&= (\mathbf{c}_{\boldsymbol{x}}^T H\boldsymbol{w} - \mathbf{h}_x^T \boldsymbol{w})^2 + \mathbf{c}_{\boldsymbol{x}}^T \sigma_Z^2 R \mathbf{c}_{\boldsymbol{x}} + \sigma_Z^2 - 2\mathbf{c}_{\boldsymbol{x}}^T \sigma_Z^2 \mathbf{r}_{\boldsymbol{x}} \\
&= \mathbf{c}_{\boldsymbol{x}}^T \sigma_Z^2 R \mathbf{c}_{\boldsymbol{x}} + \sigma_Z^2 - 2\mathbf{c}_{\boldsymbol{x}}^T \sigma_Z^2 \mathbf{r}_{\boldsymbol{x}}. \tag{5.17}
\end{aligned}$$

In order to minimize the above MSE (5.17) under the unbiasedness constraint (5.16), introducing the Lagrange multiplier $\boldsymbol{\lambda} \in \mathbb{R}^m$ and differentiating the Lagrange function with respect to $\mathbf{c}_{\boldsymbol{x}}$, we have

$$\sigma_Z^2 R \mathbf{c}_{\boldsymbol{x}} - \sigma_Z^2 \mathbf{r}_{\boldsymbol{x}} - H\boldsymbol{\lambda} = \mathbf{0}. \tag{5.18}$$

5.5 Kriging and Efficient Global Optimization

Equations (5.16) and (5.18) can be rewritten by the matrix form

$$\begin{pmatrix} 0 & H^T \\ H & \sigma_Z^2 \mathbf{R} \end{pmatrix} \begin{pmatrix} -\boldsymbol{\lambda} \\ \mathbf{c}_x \end{pmatrix} = \begin{pmatrix} \mathbf{h}_x \\ \sigma_Z^2 \mathbf{r}_x \end{pmatrix}, \tag{5.19}$$

and the best linear unbiased predictor of $Y(\boldsymbol{x})$ becomes

$$\hat{y}(\boldsymbol{x}) = \mathbf{c}_x^T \boldsymbol{y}$$
$$= (-\boldsymbol{\lambda}^T \;\; \mathbf{c}_x^T) \begin{pmatrix} 0 \\ \boldsymbol{y} \end{pmatrix}$$
$$= (\mathbf{h}_x \;\; \mathbf{r}_x)^T \begin{pmatrix} 0 & H^T \\ H & \mathbf{R} \end{pmatrix}^{-1} \begin{pmatrix} 0 \\ \boldsymbol{y} \end{pmatrix}.$$

Moreover, $\hat{y}(\boldsymbol{x})$ is represented by the form

$$\hat{y}(\boldsymbol{x}) = \mathbf{h}_x^T \hat{\boldsymbol{w}} + \mathbf{r}_x^T \mathbf{R}^{-1}(\boldsymbol{y} - H\hat{\boldsymbol{w}})$$

for the generalized least squares estimator $\hat{\boldsymbol{w}} = \left(H^T \mathbf{R}^{-1} H\right)^{-1} H^T \mathbf{R}^{-1} \boldsymbol{y}$, since

$$\begin{pmatrix} 0 & H^T \\ H & \mathbf{R} \end{pmatrix}^{-1} = \begin{pmatrix} -Q & QH^T \mathbf{R}^{-1} \\ \mathbf{R}^{-1} HQ & \mathbf{R}^{-1} - \mathbf{R}^{-1} HQH^T \mathbf{R}^{-1} \end{pmatrix},$$

where $Q = (H^T \mathbf{R}^{-1} H)^{-1}$.

Thus for the MSE of the estimate with unbiasedness, we have

$$\hat{s}^2(\boldsymbol{x}) \equiv \text{MSE}(\hat{y}(\boldsymbol{x}))$$
$$= E \left(\mathbf{c}_x^T \boldsymbol{y} - Y(\boldsymbol{x}) \right)^2$$
$$= \sigma_Z^2 \left(1 + \mathbf{c}_x^T \mathbf{R} \mathbf{c}_x - 2 \mathbf{c}_x^T \mathbf{r}_x \right)$$
$$= \sigma_Z^2 \left[1 - (\mathbf{h}_x^T, \mathbf{r}_x^T) \begin{pmatrix} 0 & H^T \\ H & \mathbf{R} \end{pmatrix}^{-1} \begin{pmatrix} \mathbf{h}_x \\ \mathbf{r}_x \end{pmatrix} \right].$$

Now, the mean and variance of prediction can be summarized as

$$\hat{y}(\boldsymbol{x}) = \mathbf{h}_x^T \hat{\boldsymbol{w}} + \mathbf{r}_x^T \mathbf{R}^{-1}(\boldsymbol{y} - H\hat{\boldsymbol{w}}), \tag{5.20}$$
$$\hat{s}^2(\boldsymbol{x}) = \hat{\sigma}_Z^2 \left[1 - (\mathbf{h}_x^T, \mathbf{r}_x^T) \begin{pmatrix} 0 & H^T \\ H & \mathbf{R} \end{pmatrix}^{-1} \begin{pmatrix} \mathbf{h}_x \\ \mathbf{r}_x \end{pmatrix} \right],$$

where $\hat{\boldsymbol{w}}$ and $\hat{\sigma}_Z^2$ are given by the formulations (5.11) and (5.12):

$$\hat{\boldsymbol{w}} = \left(H^T \mathbf{R}^{-1} H\right)^{-1} H^T \mathbf{R}^{-1} \boldsymbol{y}, \tag{5.21}$$
$$\hat{\sigma}_Z^2 = \frac{1}{\ell} (\boldsymbol{y} - H\hat{\boldsymbol{w}})^T \mathbf{R}^{-1} (\boldsymbol{y} - H\hat{\boldsymbol{w}}).$$

Parameter Estimation in Ordinary Kriging

The general form of Kriging in geostatistics is given by

$$Y(\boldsymbol{x}) = \mu(\boldsymbol{x}) + Z(\boldsymbol{x}), \tag{5.22}$$

where $\mu(\boldsymbol{x})$ is a deterministic mean structure of model and $Z(\boldsymbol{x})$ is a *stochastic process*[3] with $E(Z(\boldsymbol{x})) = 0$.

As typical types of Kriging, there are simple Kriging, ordinary Kriging and universal Kriging:

- Simple Kriging: $\mu(\boldsymbol{x})$ is known and constant
- Ordinary Kriging: $\mu(\boldsymbol{x})$ is unknown, but constant
- Universal Kriging: $\mu(\boldsymbol{x}) = \sum_{i=1}^{m} w_i h_i(\boldsymbol{x})$

In the next section, we will introduce the (ordinary) Kriging-based efficient global optimization in which Y is assumed to be normally distributed. Here, we describe the necessary parameters to be estimated in the stochastic model (5.8) as follows.

Letting a basis function $h(\boldsymbol{x}) = 1$, i.e., $Y(\boldsymbol{x}) = w + Z(\boldsymbol{x}) = \mu + Z(\boldsymbol{x})$ in the stochastic model (5.8), the mean $\hat{y}(\boldsymbol{x})$ and the variance $\hat{s}^2(\boldsymbol{x})$ of prediction at a point $\boldsymbol{x}$ can be obtained from (5.20) and (5.21) as follows:

$$\hat{y}(\boldsymbol{x}) = \hat{\mu} + \mathbf{r}_{\boldsymbol{x}}^T \mathbf{R}^{-1}(\boldsymbol{y} - \mathbf{1}\hat{\mu}), \tag{5.23}$$

$$\hat{s}^2(\boldsymbol{x}) = \hat{\sigma}_Z^2 \left(1 - \mathbf{r}_{\boldsymbol{x}}^T \mathbf{R}^{-1} \mathbf{r}_{\boldsymbol{x}} + \frac{(1 - \mathbf{1}^T \mathbf{R}^{-1} \mathbf{r}_{\boldsymbol{x}})^2}{\mathbf{1}^T \mathbf{R}^{-1} \mathbf{1}} \right), \tag{5.24}$$

where

$$\hat{\mu} = \hat{w} = \left(\mathbf{1}^T \mathbf{R}^{-1} \mathbf{1} \right)^{-1} \mathbf{1}^T \mathbf{R}^{-1} \boldsymbol{y},$$

$$\hat{\sigma}_Z^2 = \frac{1}{\ell} \left(\boldsymbol{y} - \mathbf{1}\hat{w} \right)^T \mathbf{R}^{-1} \left(\boldsymbol{y} - \mathbf{1}\hat{w} \right).$$

Note that $y(\boldsymbol{x}_i) = \hat{\mu} = y_i$ and $\hat{s}^2(\boldsymbol{x}_i) = 0$ for given sample data $(\boldsymbol{x}_i, y_i)$, $i = 1, \ldots, \ell$. One may see this as follows:

For a sample point $\boldsymbol{x}_i$, we have the following:

$$\mathbf{R}^{-1}\mathbf{r}_{\boldsymbol{x}_i} = [\mathbf{r}_{\boldsymbol{x}_1}, \ldots, \mathbf{r}_{\boldsymbol{x}_\ell}]^{-1} \mathbf{r}_{\boldsymbol{x}_i} = \boldsymbol{e}_i,$$

---

[3] In probability theory, a stochastic (or random) process is the counterpart to a deterministic process, and is a random function with respect to time. That is, a stochastic process which changes under the time has some uncertainty described by probability distributions. As examples of stochastic process, there are stock market, exchange rate fluctuations, and random movements such as Brownian motion of which *Wiener process* is well known as the mathematical model. Applying this definition, a stochastic process in geostatistics means a random function with respect to location.

5.5 Kriging and Efficient Global Optimization

where e_i is the ith unit column vector. Thus, (5.23) becomes

$$\hat{y}(\boldsymbol{x}_i) = \hat{\mu} + \mathbf{r}_{\boldsymbol{x}_i}^T \mathbf{R}^{-1}(\boldsymbol{y} - \mathbf{1}\hat{\mu})$$
$$= \hat{\mu} + \boldsymbol{e}_i^T(\boldsymbol{y} - \mathbf{1}\hat{\mu}) = \hat{\mu} + (y_i - \hat{\mu})$$
$$= y_i,$$

which implies the prediction by Kriging is an *interpolation*. Moreover, (5.24) reduces to

$$\hat{s}^2(\boldsymbol{x}_i) = 0,$$

because

$$\mathbf{r}_{\boldsymbol{x}_i}^T \mathbf{R}^{-1} \mathbf{r}_{\boldsymbol{x}_i} = \mathbf{r}_{\boldsymbol{x}_i}^T \boldsymbol{e}_i = R(\boldsymbol{x}_i, \boldsymbol{x}_i) = 1,$$
$$\mathbf{1}^T \mathbf{R}^{-1} \mathbf{r}_{\boldsymbol{x}}^2 = \mathbf{1}^T \boldsymbol{e}_i = 1.$$

Figure 5.11 shows the prediction by ordinary Kriging.

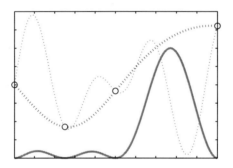

Fig. 5.11 Prediction by ordinary Kriging: the *dotted line* represents the true function, the *dashed line* does the predicted value $\hat{y}$ and the *solid line* does the variance $\hat{s}^2$ of prediction providing information of uncertainty

5.5.2 Efficient Global Optimization

In this section, we describe the *efficient global optimization* (EGO) which has been suggested and developed by [64, 65, 134, 135]. The EGO is a kind of global optimization methods using Bayesian approach (*Bayesian global optimization*) [9, 35, 81, 92, 93, 150, 166], and uses the Kriging model as a prediction method for unknown function. Especially, the EGO adopts the *expected improvement* (EI) for selecting new sample data. Originally, the idea of EI was introduced by Kushner [78].

The *Kushner's criterion* determines a next sample point by maximizing the probability to improve the value of the objective function on the current best point f_{min}^k after k function evaluations: For a given positive value of ϵ

$$\text{maximize} \quad \Pr\left(\hat{y}(\boldsymbol{x}) < f_{min}^k - \epsilon\right). \tag{5.25}$$

If a random variable Y is normally distributed, the above problem (5.25) becomes

$$\text{maximize} \quad \varPhi\left(\frac{(f_{min}^k - \epsilon) - \hat{y}(\boldsymbol{x})}{\hat{s}(\boldsymbol{x})}\right),$$

where $\varPhi(\cdot)$ is the cumulative density function of the standard normal distribution, and $\hat{y}(\boldsymbol{x})$ and $\hat{s}^2(\boldsymbol{x})$ are, respectively, the mean and the variance of prediction at the point $\boldsymbol{x}$. For simplicity, denote $\hat{y}(\boldsymbol{x})$ as $\hat{y}$ and $\hat{s}(\boldsymbol{x})$ as $\hat{s}$.

By changing the value of ϵ, we can control how globally (large ϵ) or how locally (small ϵ) the design space is searched. Conventionally, starting with a large value of ϵ for global search, we take a smaller value of ϵ for local search as the iteration proceeds.

Remark 5.8. Another criterion [92] is to minimize a (posteriori) risk function, namely the average deviation from the current best point, which is known as the one-step Bayesian method:

$$\text{minimize} \quad \frac{1}{\sqrt{2\pi}\hat{\sigma}} \int_{-\infty}^{\infty} \min\left(y, f_{min}^k - \epsilon\right) \exp\left(\frac{-(y - \hat{y})^2}{2\hat{\sigma}^2}\right) dy,$$

where $\hat{y}$ and $\hat{s}^2$ are conditional mean and variance at the point x, respectively.

Expected Improvement

Kushner's criterion depends on the parameter ϵ, thus it is important how to choose an appropriate value for ϵ. In order to overcome the difficulty for selecting an appropriate ϵ in Kushner's criterion and to search more effectively globally and locally in the design space, the criterion of *expected improvement* was proposed by Jones et al. [65].

The improvement I over the current best value f_{min}^k after k evaluations is defined by

$$I(\boldsymbol{x}) = \max\left(f_{min}^k - Y, 0\right),$$

and the *expected value of improvement (expected improvement)* can be computed as

$$E(I) \equiv E\left(\max\left(f_{min}^k - Y, 0\right)\right),$$

$$= \int_{-\infty}^{f_{min}^k} \frac{f_{min}^k - y}{\hat{s}} \phi\left(\frac{y - \hat{y}}{\hat{s}}\right) dy$$

$$= \left(f_{min}^k - \hat{y}\right)\varPhi\left(\frac{f_{min}^k - \hat{y}}{\hat{s}}\right) + \hat{s}\phi\left(\frac{f_{min}^k - \hat{y}}{\hat{s}}\right), \qquad (5.26)$$

where $\phi(\cdot)$ and $\varPhi(\cdot)$ are the density function and the cumulative distribution function of the standard normal distribution, respectively, and $\hat{y}$ and $\hat{s}$ in the Kriging-based EGO are given by (5.23) and (5.24).

The criterion is to maximize the function of the expected improvement (5.26) which can be interpreted from two viewpoints of finding an optimal

5.5 Kriging and Efficient Global Optimization

point and predicting unknown objective function. Differentiating the function of the expected improvement with respect to $\hat{y}$ and $\hat{s}$, the following holds:

$$\frac{\partial E(I)}{\partial \hat{y}} = -\Phi\left(\frac{f_{min}^k - \hat{y}}{\hat{s}}\right) < 0,$$

$$\frac{\partial E(I)}{\partial \hat{s}} = \phi\left(\frac{f_{min}^k - \hat{y}}{\hat{s}}\right) > 0.$$

This means that the expected improvement is decreasing with respect to $\hat{y}$ and increasing with respect to $\hat{s}$. In other words, the expected improvement is larger for smaller $\hat{y}$ and for larger $\hat{s}$. Note that $\hat{s}^2$ is a prediction variance which represents uncertainty for prediction. As was described above, $\hat{s} = 0$ for the existing observed values, i.e., there is no error at given the sample points, while the value of $\hat{s}$ is large in the region nearby the place where there are less number of the existing sample points. Thus, additional points should be selected from the region with a large value of the expected improvement for more exact prediction of objective function and for more improvement of optimal solution. On the other hand, the second term of expected improvement (5.26) does not appear in Kushner's criterion which does not take into account the effect of $\hat{s}$ for prediction, and hence lacks in viewpoint of more exact prediction of objective function (Fig. 5.12).

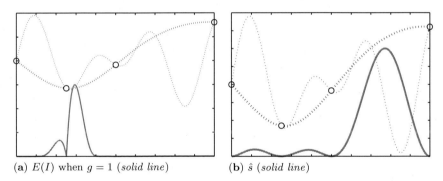

(a) $E(I)$ when $g = 1$ (*solid line*) (b) $\hat{s}$ (*solid line*)

Fig. 5.12 Expected improvement (**a**) and $\hat{s}$ (**b**): the *dotted line* represents the true function and the *dashed line* does the mean of prediction

Generalized Expected Improvement

As was stated above, the expected improvement can take account of global and local search together. For searching more globally or locally according to situations, the expected improvement was extended by introducing some parameter [134, 135].

The improvement I^g for a nonnegative parameter g is defined by

$$I^g(\boldsymbol{x}) = \max\left\{\left(f_{min}^k - Y\right)^g, 0\right\}.$$

For the cases of $g = 1, 2, \ldots$, the generalized expected improvement is expressed by

$$E(I^g) = \hat{s}^g \sum_{j=0}^{g} (-1)^j \frac{g!}{j!(g-j)!} \left(\frac{f_{min}^k - \hat{y}}{\hat{s}} \right)^{g-j} T_j, \qquad (5.27)$$

where T_j satisfies the following recursive equation (for details, see [134]):

$$T_0 = \Phi \left(\frac{f_{min}^k - \hat{y}}{\hat{s}} \right),$$

$$T_1 = -\phi \left(\frac{f_{min}^k - \hat{y}}{\hat{s}} \right),$$

$$T_j = -\phi \left(\frac{f_{min}^k - \hat{y}}{\hat{s}} \right) \left(\frac{f_{min}^k - \hat{y}}{\hat{s}} \right)^{j-1} + (j-1)T_{j-2}, \quad j \geqq 2.$$

Remark 5.9. Consider the special cases with $g = 0, 1, 2$:

- $g = 0$: the expected improvement becomes the probability of improvement

$$E(I^0) = \Phi \left(\frac{f_{min}^k - \hat{y}}{\hat{s}} \right),$$

  which was used in the algorithm by Žilinskas [150].
- $g = 1$: it can be easily shown that the generalized expected improvement (5.27) is equivalent to the (5.26).
- $g = 2$: using (5.26) and (5.27), the following holds:

$$E(I^2) = \hat{s}^2 \left[\left(\left(\frac{f_{min}^k - \hat{y}}{\hat{s}} \right)^2 + 1 \right) \Phi \left(\frac{f_{min}^k - \hat{y}}{\hat{s}} \right) + \frac{f_{min}^k - \hat{y}}{\hat{s}} \phi \left(\frac{f_{min}^k - \hat{y}}{\hat{s}} \right) \right]$$

$$= (E(I))^2 + \mathrm{Var}(I),$$

  which means that the variance (uncertainty) and the expectation of improvement can be considered together.

For example, for the cases of $g = 1, 5, 10$, the expected improvements are shown in Fig. 5.13a–c. The dotted line is the true function, the dashed line is the predicted function by the ordinary Kriging, and circles o are sample points. At the bottom, the solid line shows the generalized expected improvement.

From Fig. 5.13a with $g = 1$, the value of expected improvement is large around the current best point to the predicted function, since the value g and the number of sample points is relatively small. After adding more sample points, Fig. 5.13a' shows that the expected improvement is large at the point nearby where the prediction variance (standard error) may be large. When $g = 5$ as shown in Fig. 5.13b, the expected improvement is almost the same

5.5 Kriging and Efficient Global Optimization

with the one for $g = 1$ in this example. In the case of sufficiently large value $g = 10$, however, the expected improvement has a large value in the sparse place of sample points.

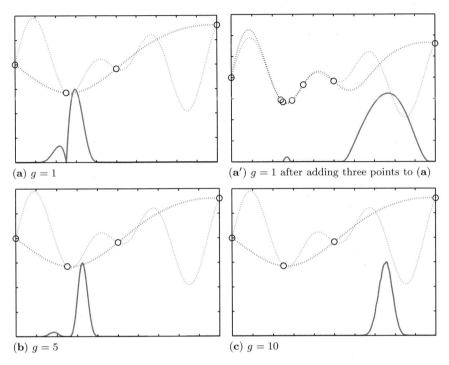

Fig. 5.13 Expected improvements for several cases

The generalized expected improvement can adjust global or local search by controlling the value of the parameter g. However, it is not easy to choose an appropriate value of g in advance, but usually the value of g is taken larger as the iteration proceeds. In our idea, it seems better to calculate simultaneously the expected improvements for two cases with small and large g, as seen in Example 5.5 below.

Example 5.5. Consider a simple example given by

$$\underset{x_1, x_2}{\text{maximize}} \quad f(\boldsymbol{x}) = 10 \exp\left(-\frac{(x_1 - 10)^2 + (x_2 - 15)^2}{100}\right) \sin x_1$$

subject to $\quad 0 \leq x_1 \leq 15, \ 0 \leq x_2 \leq 20.$

This problem has an optimal (maximal) value $f^* = 9.5585$ at $x_1^* = 7.8960$ and $x_2^* = 15$, and the true configuration and contour of the objective function are shown in Fig. 5.14.

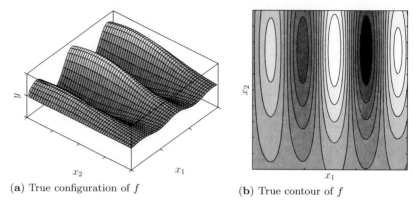

(a) True configuration of f (b) True contour of f

Fig. 5.14 Objective function in Example 5.5

The simulations are implemented along the procedure as follows:

Step 1. Predict the form $\hat{f}$ of the objective function by using the ordinary Kriging on the basis of the given sample data $(\boldsymbol{x}_i, y_i)$, $i = 1, \ldots, \ell$. Note that $y_i = f(\boldsymbol{x}_i)$.

Step 2. Find an optimal solution to the predicted objective function $\hat{f}$ by some genetic algorithm (GA) which was introduced in Sect. 3.1.

Step 3. Calculate the expected improvement $E(I^g)$ for a certain value of g which may be given as plural values.

Step 4. Terminate the iteration if the stop condition is satisfied, i.e., when the expected improvement is sufficiently small. Otherwise go to the next step.

Step 5. Choose a point $\boldsymbol{x} = \arg\max_{\boldsymbol{x}} E(I^g)$ as the next sample point, and go to Step 1.

Giving $g = 1$ for local search and $g = 10$ for global search, we start the initial iteration with five points $(0,0)$, $(15,0)$, $(0,20)$, $(7.5,10)$, $(15,20)$, namely a center point and four points at corners. After the first iteration, the result is shown in Fig. 5.15. Circles ∘ are sample points, a triangle △ is the optimal point for the predicted objective function $\hat{f}$, and squares □ are points which maximize the expected improvement for $g = 1$ and $g = 10$.

At this iteration, choosing two points with maximal expected improvement as the next sample points, we obtain the result shown in Fig. 5.16 for two values of g. After that, intermediate results are shown in Figs. 5.17–5.19, and Figure 5.20 is a final result of the simulation. Figure 5.21 shows the convergence process of the optimal value $\hat{f}^*$ for the predicted objective function and the current best value f_{min} among sampled data.

As is seen in Fig. 5.20, the predicted solution is very close to the true one, which means that the objective function may be predicted well in a neighborhood of the true optimal point. The sampled points spread relatively well over the whole of design space. However the sampled points are scarcely

5.5 Kriging and Efficient Global Optimization

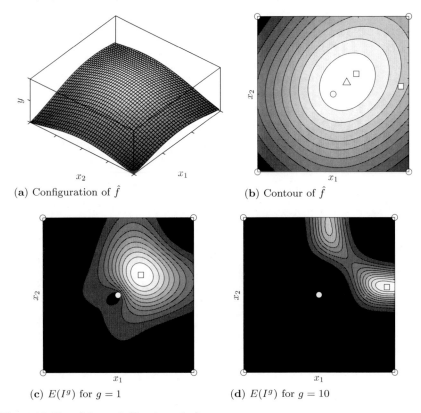

(a) Configuration of $\hat{f}$

(b) Contour of $\hat{f}$

(c) $E(I^g)$ for $g = 1$

(d) $E(I^g)$ for $g = 10$

Fig. 5.15 No. of data = 5 (first iteration)

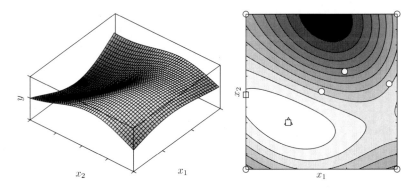

Fig. 5.16 No. of data = 7

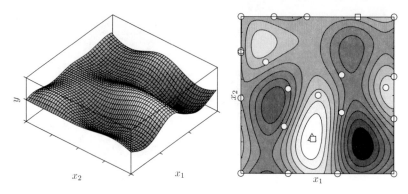

Fig. 5.17 No. of data = 21

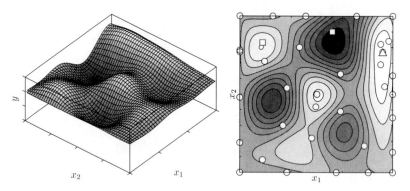

Fig. 5.18 No. of data = 35

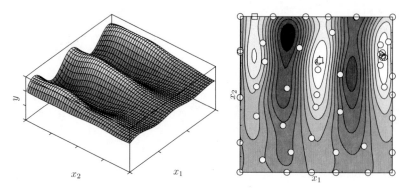

Fig. 5.19 No. of data = 49

allocated in two dark areas with lower values of objective function. This is because the objective function of Example 5.5 is to be maximized, and hence the expected improvement is low in the areas with small values of objective function.

5.6 Distance-Based Local and Global Information

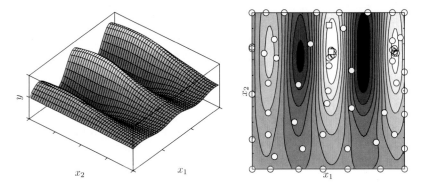

$\hat{x}_1^* = 7.8964$, $\hat{x}_2^* = 14.9819$, $\hat{f}^* = 9.5587$

Fig. 5.20 No. of data = 63 (final iteration)

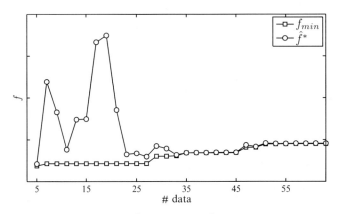

Fig. 5.21 Convergence process of $\hat{f}^*$ and f_{min} ($\hat{f}^*$: optimal value for the predicted objective function, f_{min}: the current best value among sample data)

5.6 Distance-Based Local and Global Information

In optimization problems with an objective function of unknown form, we have two problems in the followings:

- Finding an approximate optimal solution[4] to the true one as precisely as possible
- Predicting the function as accurately as possible

In order to attain our aims described above, we need two kinds of information, namely local and global. With the lack of global information for

---

[4] The optimal solution to the predicted objective function is called an approximate optimal solution.

predicting the objective function, the obtained optimal solution to the predicted objective function may converge to a local optimum. On the other hand, without considering local information, the obtained optimal solution may lose desirable precision.

Nakayama et al. [106, 107] have suggested the method which gives both global information for predicting the objective function and local information near the optimal point at the same time. In this method, two kinds of additional data are taken for relearning the form of the objective function. One of them is selected from a neighborhood of the current optimal point in order to add local information near the (estimated) optimal point. The size of this neighborhood is controlled during the convergence process. The other one is selected far away from the current optimal value and in a sparse area of existing samples in order to give a better prediction of the form of the objective function (see Fig. 5.22). The former additional data gives more detailed information near the current optimal point. The latter data prevents from converging to local optimal point.

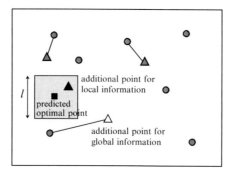

Fig. 5.22 Geometric interpretation of two additional points

The neighborhood of the current optimal point is given by a square S, whose center is the current optimal point, with the length of a side l. Let S_0 be a square, whose center is the current optimal point, with the fixed length of a side l_0. The square S is shrunk according to the number C_x of optimal points appeared continuously in S_0 in the past. That is,

$$l = l_0 \times \frac{1}{C_x + 1}. \qquad (5.28)$$

The first additional data is selected inside the square S randomly. The second additional data is selected in an area, in which the existing learning data are sparse, outside the square S. An area with sparsely existing data may be found as follows:

5.6 Distance-Based Local and Global Information

1. First, a certain number (N_{rand}) of points are generated randomly outside the square S.
2. Denote d_{ij} the distance from these points p_i $(i = 1, \ldots, N_{rand})$ to the existing training samples q_j $(j = 1, \ldots, N)$. Select the shortest k distances $\tilde{d}_{ij}$ $(j = 1, \ldots, k)$ for each p_i, and sum up these k distances, i.e., $D_i = \sum_{j=1}^{k} \tilde{d}_{ij}$.
3. Take p_t which maximizes $\{D_i\}_{(i=1,\ldots,N_{rand})}$ as an additional data outside S.

The algorithm using distance-based local and global information by [106] can be summarized as follows:

Step 1. Predict the form of the objective function by RBFN/SVR on the basis of the given training data.

Step 2. Find an optimal point for the predicted objective function by GA.

Step 3. Count the number of optimal points appeared continuously in the past in S_0. This number is represented by C_x.

Step 4. Terminate the iteration if either of the following conditions holds;
- If C_x is larger than or equal to the given C_x^0 a priori
- If the best value of the objective function obtained so far is identical during the last certain number (C_f^0) of iterations

Otherwise calculate l by (5.28), and go to the next step.

Step 5. Select an additional data near the current optimal value, i.e., inside S.

Step 6. Select another additional points outside S in a place in which the density of the existing training points is low as stated above.

Step 7. Go to Step 1.

For Example 5.5, we set $\lambda = 0.01$ in RBFN, population in GA $= 100$, generation in GA $= 100$, $l_0 = 4.0$, $C_f^0 = C_x^0 = 20$, $N_{rand} = 50$, and $k = 1$. The value of r is decided by the formula of simple estimate described in Sect. 4.2:

$$ r = \frac{d_{max}}{\sqrt[n]{nm}}, $$

where d_{max} is the maximal distance among the data, n is the dimension of the input data, m is the number of basis functions.

Table 5.1 for ten trials by the method using distance-based local and global information shows the results and the error which is the distance between the true optimal solution/value and the solution/value to the predicted objective function.

The intermediate iterations for the trial no. 4 in Table 5.1 are shown in Figs. 5.23–5.27, and the convergence procedure in Fig. 5.28. After 28 iterations, the method using distance-based local and global information [106] provides a good prediction of function and optimal solution as shown in Fig. 5.27. The results are similar to the ones by using the expected improvement in which a more load for calculating the values of $E(I^g)$ is imposed.

Table 5.1 Results by distance-based local and global information in Example 5.5

Trial	No. of data	x_1	x_2	$\hat{f}^*$	$\boldsymbol{x}$ error	f error	f error (%)
1	81	7.888	17.181	9.0028	2.181	0.5557	5.8135
2	63	14.137	14.853	7.7716	6.243	1.7869	18.6945
3	45	8.084	13.506	8.6731	1.506	0.8854	9.2625
4	61	7.912	15.512	9.2880	0.513	0.2705	2.8303
5	95	7.832	14.537	9.5251	0.467	0.0334	0.3493
6	75	7.854	16.276	9.5013	1.276	0.0572	0.5981
7	71	8.024	14.453	9.3649	0.561	0.1936	2.0255
8	51	7.723	17.535	8.4460	2.541	1.1125	11.6386
9	59	14.012	14.881	8.2435	6.117	1.3150	13.7577
10	65	7.798	14.242	9.6906	0.764	0.1321	1.3820
Average	66.6			8.9507	2.217	0.6342	6.6352
STD	13.9			0.6075	2.091	0.5799	6.0667

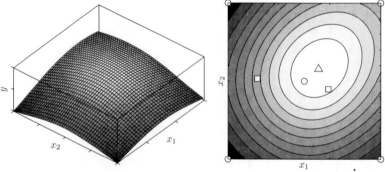

Fig. 5.23 No. of data = 5 (initial iteration)

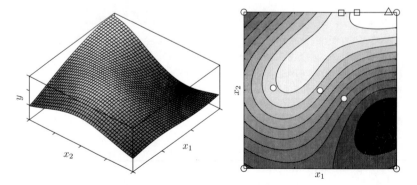

Fig. 5.24 No. of data = 7

5.6 Distance-Based Local and Global Information　　　　　　　　　　　　145

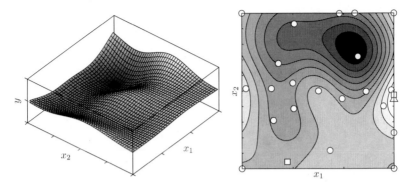

Fig. 5.25 No. of data = 21

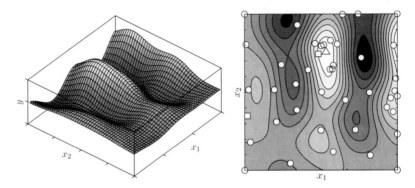

Fig. 5.26 No. of data = 41

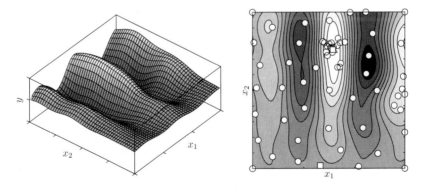

$\hat{x}_1^* = 7.8947,\ \hat{x}_2^* = 15.1424,\ \hat{f}^* = 9.2880$

Fig. 5.27 No. of data = 61 (final iteration)

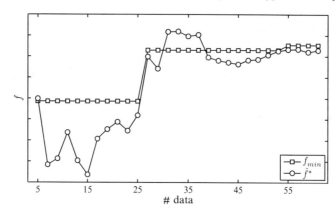

Fig. 5.28 Convergence process of $\hat{f}^*$ and f_{min} ($\hat{f}^*$: optimal value for the predicted objective function, f_{min}: the current best value among sample data)

Pressure Vessel Problem

Applying this method to a more practical problem, consider the pressure vessel design problem given by Sandgren [127], Kannan and Kramer [68], Qian et al. [118], Hsu et al. [60], Lin et al. [80], Kitayama et al. [69].

Example 5.6. As shown in Fig. 5.29, the design variables are R and L, the inner radius and length of the cylindrical selection, and Ts and Th are integer multiples of 0.0625 (in.), the available thickness of rolled steel plates.

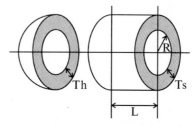

Fig. 5.29 Pressure vessel

The objective is to minimize the total cost of manufacturing the pressure vessel, including the cost of material and cost of forming and welding. The constraints g_1 and g_2 correspond to ASME limits on the geometry while g_3 corresponds to a minimum volume limit. So, the problem can be summarized as follows:

Design variables:

  R: inner radius (continuous)
  L: length of the cylindrical section (continuous)
  Ts: thickness of rolled steel plate (discrete)
  Th: thickness of rolled steel plate (discrete)

5.6 Distance-Based Local and Global Information

Constraints:

g_1: minimum shell wall thickness

$$\frac{0.0193R}{Ts} \leqq 1.0$$

g_2: minimum head wall thickness

$$\frac{0.00954R}{Th} \leqq 1.0$$

g_3: minimum volume of tank

$$\frac{1{,}296{,}000 - \frac{4}{3}\pi R^3}{\pi R^2 L} \leqq 1.0$$

Side constraints:

$$10 \leqq R \leqq 200$$
$$10 \leqq L \leqq 200$$
$$0.0625 \leqq Ts \leqq 6.1875$$
$$0.0625 \leqq Th \leqq 6.1875$$

Objective function:

$$f = 0.6224TsRL + 1.7781ThR^2 + 3.1661Ts^2L + 19.84Ts^2R$$
$$\longrightarrow \text{minimize} \tag{5.29}$$

In this problem, the objective function is explicitly given in terms of design variables. However, many researchers use this problem as a benchmark, and hence we use the problem in order to see the effectiveness of the stated method using distance-based local and global information by treating it in such a way that the objective function is not known. The parameters of the RBFN are given by $\lambda = 0.0001$, and the radius r is decided by (4.7). In this example, we use also GA: all variables are transformed into binary variables of 10 bit, after then the round off procedure[5] is applied to discrete variables. The population is 100, the generation 100, and the rate of mutation 10%. The constraint functions are penalized by

$$F(\boldsymbol{x}) = f(\boldsymbol{x}) + \sum_{j=1}^{3} p_j \times [P(g_j(\boldsymbol{x}))]^a,$$

---

[5] Let a decimal scale d according to a binary b, and $x_1 \leqq x_2 \leqq \ldots \leqq x_N$ be the values which a discrete variable can take. Then, find the integer I nearest to d/N, and the value of real code is given by Ith value, i.e., x_I.

where we set $p_j = 100$ (penalty coefficient), $a = 2$ (penalty exponent) and $P(y) = \max\{y, 0\}$. The original objective function (5.29) can be converted to minimizing the augmented objective function $F(\boldsymbol{x})$.

The data for initial training is 17 points at the corners of the region given by the side constraints and its center. Finally, each side of the rectangle controlling the convergence and the neighborhood of the point under consideration is half the upper bound–the lower bound of each side constraint. The results of simulation for the stop condition $C_x^0 = 50$ and $C_f^0 = 30$ are given in the following table.

Figure 5.30 shows the convergence process in the trial no. 4 of Table 5.2, where f_{min} represents the best value obtained so far, while $\hat{f}^*$ is the optimal value for the predicted objective function.

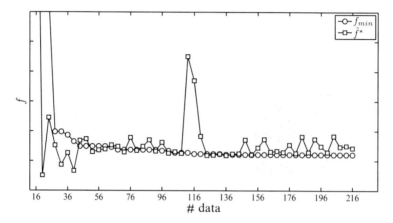

Fig. 5.30 Convergence process of example (5.29)

Table 5.2 Simulation results of example (5.29) with $C_x^0 = 50$ and $C_f^0 = 30$

Trial	No. of data	R (in.)	L (in.)	Ts (in.)	Th (in.)	f_{min} ($)
1	170	45.25	141.00	0.8750	0.4375	6,096.69
2	226	42.06	177.07	0.8125	0.4375	6,062.55
3	160	44.52	152.40	0.8750	0.4375	6,282.18
4	216	42.13	176.20	0.8125	0.4375	6,055.57
5	188	44.83	153.78	0.8750	0.4375	6,371.60
6	188	42.03	181.02	0.8125	0.4375	6,150.73
7	188	45.04	144.04	0.8750	0.4375	6,144.72
8	190	41.91	183.80	0.8125	0.4375	6,195.68
9	156	42.07	177.10	0.8125	0.4375	6,065.97
10	180	48.51	110.67	0.9375	0.5000	6,379.07
average	186.20					6,180.47
STD	20.97					117.62

5.6 Distance-Based Local and Global Information

Table 5.3 Comparison among several methods

	R (in.)	L (in.)	Ts (in.)	Th (in.)	f_{min} ($)
Sandgren	47.00	117.70	1.1250	0.6250	8,129.80
Qian	58.31	44.52	1.1250	0.6250	7,238.83
Kannan	58.29	43.69	1.1250	0.6250	7,198.20
Lin	NA	NA	NA	NA	7,197.70
Hsu	NA	NA	NA	NA	7,021.67
He	42.10	176.64	0.8125	0.4375	6,059.71
Kitayama	42.37	173.42	0.8125	0.4375	6,029.87
Nakayama	42.13	176.20	0.8125	0.4375	6,055.57

Table 5.3 shows the comparison among the existing methods. In this table, the best solution by each method is listed (Sandgren [127], Qian et al. [118], Kannan and Kramer [68], Lin et al. [80], Hsu et al. [60], He et al. [56], and Kitayama et al. [69]). Sandgren [127] and Kannan and Kramer [68] use gradient type optimization techniques which treat constraints by penalty function or augmented Lagrangian methods. Lin et al. [80] uses simulated annealing method and genetic algorithm, Qian et al. [118] uses genetic algorithms. He et al. [56] and Kitayama et al. [69] use particle swarm optimizations. The existing methods do not use metamodels, thus they needed a large number of function evaluations is too large, say more than 10,000. On the other hand, it should be noted that the method using metamodels on the basis of distance-based local and global information [106] can provide good results with only about 200 sampled points.

Chapter 6
Combining Aspiration Level Approach and SAMO[1]

In multiobjective optimization, it is one of main issues how to obtain Pareto optimal solutions, and how to choose one solution from many Pareto optimal solutions. To this end, *interactive optimization methods* [47, 90, 130, 147, 157, 159], e.g., *aspiration level method* [100] introduced in Chap. 2, have been developed. Aspiration level method searches a final solution by processing the following two stages repeatedly (1) solving auxiliary optimization problem to obtain the closest Pareto optimal solution to a given aspiration level of decision maker and (2) revising her/his aspiration level by making the trade-off analysis. Conventional interactive optimization methods are useful in particular in cases with many objective functions, in which it is difficult to visualize Pareto frontier, and also to depict the trade-off among many objective functions.

In cases with two or three objective functions, on the other hand, it may be the best way to depict Pareto frontier, because visualizing Pareto frontier helps to grasp the trade-off among objective functions. For that purpose, evolutionary methods such as genetic algorithm (GA) have been effectively applied for solving multiobjective optimization problems: so-called *evolutionary multiobjective optimization* (EMO) methods have been proposed for generating Pareto optimal solutions [24, 30, 31, 44, 49, 59, 72, 131, 144, 169, 170]. However, EMO has some problems (1) it is difficult to visualize Pareto frontiers in cases with many objective functions and (2) many function evaluations are usually needed in generating the whole Pareto frontier. Considering the number of function evaluations, it would be rather reasonable to generate not the whole Pareto frontier, but a necessary part of it in which the decision maker may be interested. To this aim, we introduce some methods combining aspiration level approach and computational intelligence method in this chapter.

---

[1] Sequential Approximate Multiobjective Optimization.

H. Nakayama et al., *Sequential Approximate Multiobjective Optimization Using Computational Intelligence*, Vector Optimization, DOI 10.1007/978-3-540-88910-6_6, © 2009 Springer-Verlag Berlin Heidelberg

151

6.1 Sequential Approximate Multiobjective Optimization Using Satisficing Trade-off Method

We write again a multiobjective optimization problem (MOP) below:

$$\underset{\boldsymbol{x}}{\text{minimize}} \quad \boldsymbol{f}(\boldsymbol{x}) = (f_1(\boldsymbol{x}), \dots, f_r(\boldsymbol{x}))^T \qquad \text{(MOP)}$$

$$\text{subject to} \quad \boldsymbol{x} \in X = \{ \, \boldsymbol{x} \in \mathbb{R}^n \mid g_j(\boldsymbol{x}) \leqq 0, \; j = 1, \dots, m \, \},$$

where X denotes the set of all feasible solutions in the design variable space.

To begin with, we summarize the method for *sequential approximate multiobjective optimization* (SAMO) using satisficing trade-off method proposed by Yoon et al. [163] as follows:

Step 1. Calculate the real values of objective functions $\boldsymbol{f}(\boldsymbol{x}_1), \dots, \boldsymbol{f}(\boldsymbol{x}_\ell)$ for given initial points $\boldsymbol{x}_1, \dots, \boldsymbol{x}_\ell$.

Step 2. Approximate each objective function $f_k(\boldsymbol{x})$, $k = 1, \dots, r$, by using SVR/RBFN on the basis of training data set $(\boldsymbol{x}_i, f_k(\boldsymbol{x}_i))$, $i = 1, \dots, \ell$. Here, an *approximate function* for $f_k(\boldsymbol{x})$ is denoted by $\hat{f}_k(\boldsymbol{x})$, and an optimal solution/value to approximate objective function $\hat{\boldsymbol{f}}(\boldsymbol{x})$ is called an *approximate optimal solution/value*.

Step 3. Find an approximate optimal solution $\boldsymbol{x}^a$ closest to the given aspiration level $\overline{\boldsymbol{f}}$ by solving the following problem (AP) of satisficing trade-off method:

$$\underset{\boldsymbol{x}, \, z}{\text{minimize}} \quad z + \alpha \sum_{i=1}^{r} w_i \hat{f}_i(\boldsymbol{x}) \qquad \text{(AP)}$$

$$\text{subject to} \quad w_i \left(\hat{f}_i(\boldsymbol{x}) - \overline{f}_i \right) \leqq z, \; i = 1, \dots, r,$$

$$\boldsymbol{x} \in X,$$

where α is a sufficiently small positive number, e.g., 10^{-6}, $w_i = 1/(\overline{f}_i - f_i^*)$ and f_i^* is an ideal value.

Furthermore, generate Pareto optimal solutions $\boldsymbol{x}^1, \dots, \boldsymbol{x}^p$ to approximate objective functions $\hat{\boldsymbol{f}}$ by using EMO for finding the whole set of Pareto solutions and for deciding the second additional samples which will be stated in Step 5 later.

Step 4. Stop the iteration if a certain stop condition is satisfied. Otherwise, go to the next step. The stop condition is given by, e.g., the limitation of the number of sample points, the count of no-changing approximate solution obtained in Step 3, and so on.

Step 5. Choose additional sample points for relearning, and go to Step 1.

It is important to improve the prediction ability for function approximation in order to find an approximate solution closer to the exact one with as small number of sample data as possible. To this aim, starting with relatively few initial samples, we add new samples step by step, if necessary. Here, we introduce one of the ways how to choose additional sample points proposed by Yun–Nakayama–Yoon [163]:

1. First, one additional sample point is added as the solution x^a closest to the given aspiration level which is found in Step 3. This is for approximating well a neighborhood of Pareto optimal solution closest to the aspiration level, which enables to make easily the trade-off analysis among objective functions. In this section, the additional point x^a is considered as a *local information of Pareto frontier*, because x^a can provide the information around the closest Pareto optimal solution to the aspiration level.

2. Another additional sample point is for depicting the configuration of Pareto frontier. This is for giving a rough information of the whole Pareto frontier, and we call this a *global information of Pareto frontier* in contrast with the above local information.

Stage 1. Evaluate the rank R_i for each sample point x_i, $i = 1, \ldots, \ell$, by the ranking method of Goldberg [49] (see Fig. 3.3 stated in Sect. 3.1).

Stage 2. Approximate a ranking function $\hat{R}(x)$ on the basis of training data set (x_i, R_i), $i = 1, \ldots, \ell$, by SVR/RBFN.

Stage 3. Calculate the values of ranking function $\hat{R}(x)$ for the approximate Pareto optimal solutions x^j, $j = 1, \ldots, p$, generated in Step 3.

Stage 4. Among them, select a point with the best (minimal) rank

$$x^b = \arg \min_{j=1,\ldots,p} \hat{R}(x^j).$$

For illustrating the method described in the above, first, we show results of a simple example with one design variable and two objective functions. Figures 6.1a and 6.2a show approximate objective functions on the basis of some real sample data. Figures 6.1b and 6.2b show the approximate solution closest to the given aspiration level. Finally, Figures 6.1c and 6.2c represent the approximate Pareto optimal values.

Example 6.1 (Case 1). Consider the following problem with two design variables and two objective functions:

$$\underset{x_1, x_2}{\text{minimize}} \qquad f_1(x) = x_1$$

$$f_2(x) = 1 + x_2^2 - x_1 - 0.1 \sin(5\pi x_1)$$

$$\text{subject to} \qquad 0 \leq x_1 \leq 1, \ -2 \leq x_2 \leq 2.$$

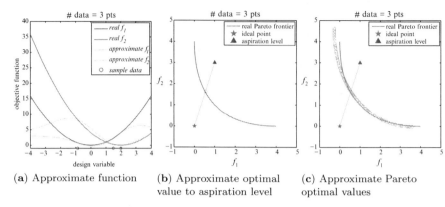

Fig. 6.1 Result with three points

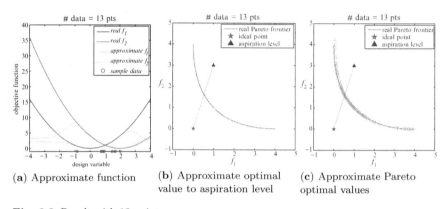

Fig. 6.2 Result with 13 points

Figure 6.3 shows the true contour of each objective function in Example 6.1.

Starting with initial sample ten points generated randomly, we stop the iteration after 15 additional learning. We approximate each objective function by using ν_ε-SVR, which is described in Sect. 4.4, with Gaussian kernel function. At the initial iteration as shown from Fig. 6.4, the approximate solution is different from the exact one, because the function approximation is not made sufficiently. However, in Fig. 6.5 of the results with 40 sample points after 15 additional learning, it is seen that the obtained approximate Pareto optimal solutions are almost the same as the exact ones.

Example 6.2. Next, we consider an engineering design problem as shown in Fig. 6.6, and the problem is formulated as follows:

$$\minimize_{h,l,t,b} \quad f_1 := 1.10471h^2 l + 0.04811 tb(14 + l)$$

$$f_2 := \frac{2.1952}{t^3 b}$$

6.1 SAMP Using Satisficing Trade-off Method

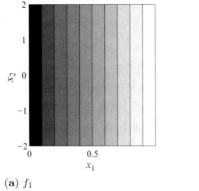
(a) f_1

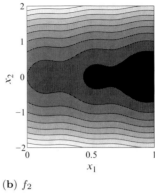
(b) f_2

Fig. 6.3 Real contours of Example 6.1

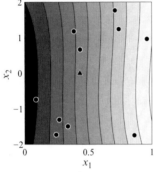
(a) Contour of $\hat{f}_1$

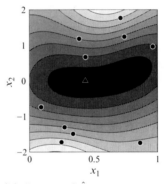
(b) Contour of $\hat{f}_2$

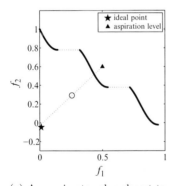
(c) Approximate value closest to aspiration level

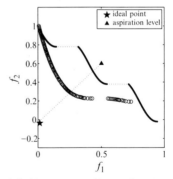
(d) Approximate Pareto frontier

Fig. 6.4 No. of data = 10 (initial iteration)

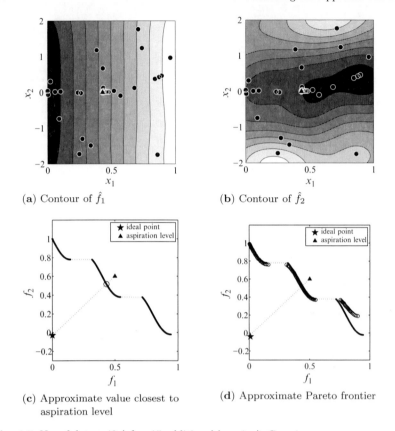

Fig. 6.5 No. of data = 40 (after 15 additional learning): Case 1

$$\text{subject to} \quad g_1 := \tau \leqq 13{,}600,$$
$$g_2 := \sigma \leqq 30{,}000,$$
$$g_3 := h - b \leqq 0,$$
$$g_4 := P_c \geqq 6{,}000,$$
$$0.125 \leqq h, \ b \leqq 5.0, \ 0.1 \leqq l, \ t \leqq 10.0,$$

where

$$\tau = \sqrt{(\tau')^2 + (\tau'')^2 + \frac{l\tau'\tau''}{\sqrt{0.25(l^2 + (h+t)^2)}}},$$

$$\tau' = \frac{6{,}000}{\sqrt{2}hl},$$

$$\tau'' = \frac{6{,}000(14 + 0.5l)\sqrt{0.25(l^2 + (h+t)^2)}}{\sqrt{2}hl\left(\frac{l^2}{12} + 0.25(h+t)^2\right)},$$

$$\sigma = \frac{504{,}000}{t^2 b}, \quad P_c = 64{,}746.022(1 - 0.0282346t)tb^3.$$

6.1 SAMP Using Satisficing Trade-off Method

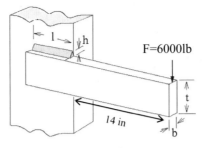

Fig. 6.6 Welded beam design problem

The ideal value and aspiration levels are given by the following:

$$\text{ideal value} := (f_1^*, f_2^*) = (0, 0)$$
$$\text{aspiration level 1} := (\overline{f}_1^1, \overline{f}_2^1) = (4, 0.003)$$
$$\text{aspiration level 2} := (\overline{f}_1^2, \overline{f}_2^2) = (20, 0.002)$$
$$\text{aspiration level 3} := (\overline{f}_1^3, \overline{f}_2^3) = (40, 0.0002)$$

Table 6.1 shows the result by the simple satisficing trade-off method using SQP with a quasi-Newton method for randomly chosen starting points in ten times. Table 6.2 shows the result by the described method [163] with 100 sample points (including 50 initial data). Figures 6.7 and 6.8 show the

Table 6.1 Result by SQP using a quasi-Newton method for real objective functions

		h	l	t	b	f_1	f_2	No. of evaluation
Asp.	Average	0.5697	1.7349	10	0.5804	5.0102	3.78E-03	249.9
level	STD	0.0409	0.1826	0	0.0072	0.0420	4.83E-05	69.6
1	Max	0.5826	2.2546	10	0.5826	5.0235	3.92E-03	369.0
	Min	0.4533	1.6772	10	0.5599	4.8905	3.77E-03	164.0
Asp.	Average	1.0834	0.8710	10.0000	1.7685	13.7068	1.25E-03	204.2
level	STD	0.3274	0.1662	5.11E-08	0.1828	1.3793	1.13E-04	30.1
2	Max	2.0132	0.9896	10	2.1263	16.3832	1.31E-03	263.0
	Min	0.9221	0.4026	10.0000	1.6818	13.0527	1.03E-03	172.0
Asp.	Average	1.7345	0.4790	10	5	36.4212	4.39E-04	251.9
level	STD	0.0000	0.0000	0	0	0.0000	5.71E-20	146.2
3	Max	1.7345	0.4790	10	5	36.4212	4.39E-04	594.0
	Min	1.7345	0.4790	10	5	36.4212	4.39E-04	112.0

Pareto frontiers by SAMO and EMO, respectively. Since we use the analytical gradient in using SQP, the number of function evaluation would be almost four times if numerical differentiation is applied for black box functions.

Table 6.2 Result by SAMO with 100 function evaluations

		h	l	t	b	f_1	f_2
Asp. level 1	Average	0.5223	1.9217	9.9934	0.5825	5.0344	3.78E-03
	STD	0.0374	0.1656	0.0136	0.0011	0.0130	1.08E-05
	Max	0.5832	2.2742	10	0.5845	5.0692	3.81E-03
	Min	0.4520	1.6859	9.9558	0.5817	5.0224	3.77E-03
Asp. level 2	Average	0.8921	1.0398	9.9989	1.6809	13.0653	1.31E-03
	STD	0.0898	0.1106	0.0012	0.0012	0.0081	7.79E-07
	Max	1.0787	1.1895	10	1.6824	13.0781	1.31E-03
	Min	0.7849	0.8273	9.9964	1.6789	13.0531	1.31E-03
Asp. level 3	Average	2.2090	0.4486	10	5	36.6830	4.39E-04
	STD	0.9355	0.2293	0	0	0.2695	5.71E-20
	Max	3.7812	0.8734	10	5	37.1257	4.39E-04
	Min	1.0391	0.1895	10	5	36.4212	4.39E-04

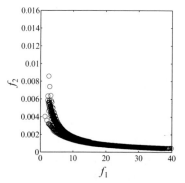
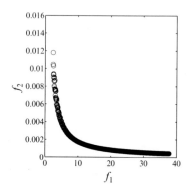
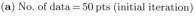

(a) No. of data = 50 pts (initial iteration)

(b) No. of data = 70 pts (after ten additional learning)

Fig. 6.7 Approximate Pareto frontier by SAMO

The stated method in the above takes into account a selection method of additional sample points to improve the prediction ability for Pareto frontier. In order to approximate well objective functions, some points should be needed in a sparse region of the design space from a viewpoint of space filling. Thus, here we can consider an additional point which is selected by distance based optimality (Sect. 5.4).

6.2 MCDM with Aspiration Level Method and GDEA

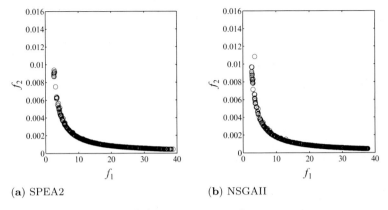

(a) SPEA2 (b) NSGAII

Fig. 6.8 Pareto frontier by EMO (100 gen. × 50 pop.)

Example 6.3 (Case 2). Note that the second additional sample at Step 5 seems to provide a local information in the sense that it is for generating a better approximation of the whole Pareto frontier. Indeed, as can be seen in Fig. 6.5, the additional samples in the space of design variables x are concentrated in the neighborhood of Pareto optimal solutions. From a viewpoint of generating a better approximation of each objective function, further sample points for global information may be needed. As the third additional sample point, we recommend to add a point in a sparse area of existing sampled points. This is performed by a similar way as the distance based local and global information method. Figure 6.9 shows the result obtained by this method. One may see that each objective function is approximated well enough and that a well-approximated Pareto frontier is generated.

The most prominent feature in SAMO described above is that combining the aspiration level method and EMO, it is possible to find the most interesting part of Pareto frontier for the decision maker as well as to grasp the configuration of the whole Pareto frontier. Furthermore, employing the approximation of objective functions, it is expected to reduce the number of function evaluations up to less than 1/100 to 1/5 of using only EMO (compare the results shown in Figs. 6.7 and 6.8).

6.2 MCDM with Aspiration Level Method and GDEA

In order to find not the whole Pareto frontier but the most interesting part of it for decision makers, Yun–Nakayama–Arakawa [161] proposed the method of combining an aspiration level approach and GDEA method explained in Sect. 3.3. As stated in Sect. 3.3, the fitness in GDEA method is evaluated by the optimal value Δ^* to the problem (GDEA$_{fit}$), which represents a relative

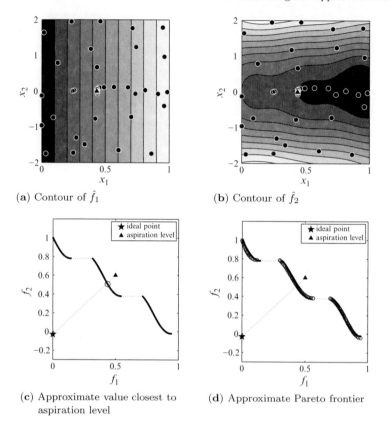

Fig. 6.9 No. of data = 40 (after ten additional learning): Case 2

degree how close to GDEA-efficient frontier. Through several examples, it has been observed that GDEA method has good convergence to Pareto frontier by changing a value of the parameter α, i.e., taking a large α can remove individuals which are located far from the Pareto frontier, and also taking a small α can generate nonconvex Pareto frontiers (see Sect. 3.3 for details on GDEA method).

For the given aspiration level of decision maker $\overline{\bm{f}} = \left(\overline{f}_1, \ldots, \overline{f}_r\right)^T$ and an ideal point $\bm{f}^* = (f_1^*, \ldots, f_r^*)^T$, the fitness for an individual $\bm{x}^o$ is evaluated by solving the problem (ALGDEA$_{fit}$)

$$\begin{aligned}
\underset{\Delta,\,\nu}{\text{maximize}} \quad & \Delta - \lambda H\left(W(\bm{f}(\bm{x}^o) - \overline{\bm{f}})\right) && (\text{ALGDEA}_{fit}) \\
\text{subject to} \quad & \Delta \leq \tilde{d}_j - \alpha \sum_{i=1}^{r} \nu_i (f_i(\bm{x}^o) - f_i(\bm{x}^j)), \ j = 1, \ldots, p, \\
& \sum_{i=1}^{r} \nu_i = 1, \\
& \nu_i \geqq \varepsilon, \ i = 1, \ldots, r, \ (\varepsilon : \text{sufficiently small number})
\end{aligned}$$

6.2 MCDM with Aspiration Level Method and GDEA

where $H(\boldsymbol{y}) = \max\{y_1,\ldots,y_r\}$, $\boldsymbol{y} = (y_1,\ldots,y_r)^T$, the matrix W is diagonal with its elements $w_i = 1/(\overline{f}_i - f_i^*)$, $i = 1,\ldots,r$, and λ is an appropriately given positive number. For instance, λ is taken as a large number when we search as close Pareto optimal values to the given aspiration level as possible.

As a result, the optimal value to the problem (ALGDEA$_{fit}$) represents the degree how close $\boldsymbol{f}(\boldsymbol{x}^o)$ is to Pareto frontier and to the given aspiration level. For example, as shown in Fig. 6.10, let $\overline{\boldsymbol{f}}$ be a given aspiration level and $\boldsymbol{f}^*$ be an ideal point. Then in the fitness assignment by (ALGDEA$_{fit}$), individuals in the neighborhood of the Pareto optimal value $A0$ have good fitness, and the Pareto optimal region shadowed in Fig. 6.10 is considered to be the most interesting part of decision makers for a final decision which is described below in more detail.

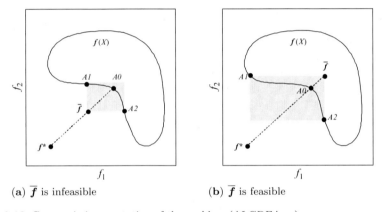

(a) $\overline{\boldsymbol{f}}$ is infeasible   (b) $\overline{\boldsymbol{f}}$ is feasible

Fig. 6.10 Geometric interpretation of the problem (ALGDEA$_{fit}$)

Among Pareto optimal solutions in the region, representative alternatives are selected by the following measures:

$$A0 = \boldsymbol{f}(\hat{\boldsymbol{x}}^0), \text{ where } \hat{\boldsymbol{x}}^0 = \arg\min_{\boldsymbol{x}^j} \max_{i=1,\ldots,r} w_i\left(f_i(\boldsymbol{x}^j) - \overline{f}_i\right), \quad (6.1)$$

$$Ai = \boldsymbol{f}(\hat{\boldsymbol{x}}^i), \text{ where } \hat{\boldsymbol{x}}^i = \arg\min_{\boldsymbol{x}^j} w_i\left(f_i(\boldsymbol{x}^j) - \overline{f}_i\right), \quad i = 1,\ldots,r, \quad (6.2)$$

where $\boldsymbol{x}^j$ is of the set of generated Pareto optimal solutions. Note that $\hat{\boldsymbol{x}}^0$ can be approximately given as the one which gives the maximum among optimal values to the problem (ALGDEA$_{fit}$) for each $\boldsymbol{x}^j$.

Figure 6.10 illustrates the meaning of the alternative $A0$ given by (6.1) and the alternatives $A1$ and $A2$ given by (6.2). When the aspiration level $\overline{\boldsymbol{f}}$ is infeasible, as seen in Fig. 6.10a, the nonnegative values of $w_i\left(f_i(\boldsymbol{x}^j) - \overline{f}_i\right)$, $i = 1,\ldots,r$, mean the degree of regret for $\boldsymbol{x}^j$. Then, the optimal solution to the problem (ALGDEA$_{fit}$) can be regarded as a compromise solution. If $\overline{\boldsymbol{f}}$ is feasible, as is in Fig. 6.10b, the nonpositive values of $w_i\left(f_i(\boldsymbol{x}^j) - \overline{f}_i\right)$,

$i = 1, \ldots, r$, mean the degree of satisfaction for $\boldsymbol{x}^j$. Thus, the obtained optimal value by solving (ALGDEA$_{fit}$) is a satisfactory solution.

Example 6.4. For illustration of the stated method, consider a simple example with two-objective functions as follows (Example 3.4 in Sect. 3.4):

$$\underset{x_1, x_2}{\text{minimize}} \quad (f_1(\boldsymbol{x}), f_2(\boldsymbol{x})) = (x_1, x_2)$$
$$\text{subject to} \quad x_1^3 - 3x_1 - x_2 \leq 0,$$
$$x_1 \geq -1, \ x_2 \leq 2.$$

To obtain Pareto optimal solutions for Example 6.4, we set parameters as follows; the size of population is 80, the probability of mutation is 0.05, $\alpha(t) = 10 \times \exp(-0.2 \times t)$, $t = 0, \ldots, 29$, $\lambda = 10^2$ and $\varepsilon = 10^{-6}$, and the aspiration level is $\overline{\boldsymbol{f}} = (0.5, 1)^T$. Given the aspiration level, we perform the above procedure until 30 generation. Then, the results are shown in Fig. 6.11. In this figure, the dotted line represents the true Pareto frontier, and the ideal point in this example is $\boldsymbol{f}^* = (-1, -2)^T$.

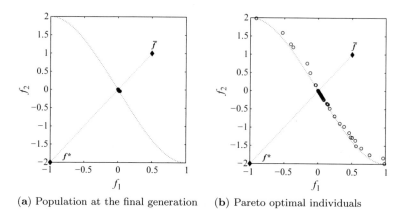

(a) Population at the final generation   (b) Pareto optimal individuals

Fig. 6.11 Result for Example 6.4

As can be seen in Fig. 6.11a, almost of all individuals at the final generation are concentrated in the neighborhood of intersection of Pareto frontier and the line segment between $\overline{\boldsymbol{f}}$ and $\boldsymbol{f}^*$. The obtained Pareto optimal individuals are candidates for final decision making solution with respect to the current aspiration level, and hence decision makers choose one from them.

Next, we apply the method to a robust design (welded beam optimization problem), which is cited from [120, 122] and reformulated on the basis of Arakawa et al.'s Fuzzy numbers method [3]. As was shown by Rao [122] and Arakawa et al. [3], the robust design can be formulated as a multiobjective optimization problem with more than three objective functions. They

6.2 MCDM with Aspiration Level Method and GDEA

used the weighted sum method to generate Pareto optimal solutions, but did not perform a trade-off analysis. Chen et al. [20] showed that the set of Pareto optimal solutions is nonconvex and utilized the weighted Tchebycheff method instead of the weighted sum method. Their method aims to explore some Pareto optimal solutions for the given aspiration level, but not to grasp trade-off among objective functions. Since the trade-off analysis is an important task in multiobjective optimization, the above methods do not work sufficiently. Through this example, we show how the described method can overcome the difficulty rising from the existing methods.

Example 6.5 (Welded Beam Design Problem). The cantilever beam, shown in Fig. 6.12, is supported at the end load $P(=26,\!689N)$ at the tip other end with its extended length $L(=0.3556m)$. The depth of the weld $h(m)$, the length of the weld $l(m)$, the height $t(m)$ and thickness $b(m)$ of the beam, the error of the weld's depth $h^R(m)$ and the error of the weld's length $l^R(m)$ are treated as design variables.

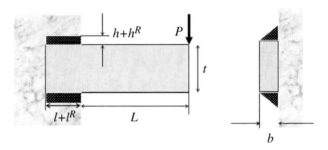

Fig. 6.12 Welded beam design

The constraints are the shearing stress g_1, the stress g_2, the deviation g_3 and the buckling load g_4. Under these conditions, the problem is to minimize the cost welding the beam f_1, to maximize the robustness with respect to the cost f_2, to maximize the robustness with respect to the fluctuation of the beam's length f_3, to maximize the error with respect to the weld's depth f_4 and to maximize the error with respect to the weld's length f_5. Then, we can summarize the problem as follows:

Design variables:

$h(m),\ l(m),\ l(m),\ t(m),\ b(m),\ h^R(m),\ l^R(m)$

Constraints:

$g_1 := \quad \tau \leq \tau_a$
$g_2 := \quad \sigma \leq \sigma_a$
$g_3 := \quad \delta \leq \delta_a$

$$g_4 := \quad P \leqq P_c$$
$$g_5 := \quad h^R + h < b$$
$$g_6 := \quad h^R \leq h$$
$$g_7 := \quad l^R \leqq l$$

Objective functions:

minimize $\quad f_1 := 68{,}216.1h^2l + 2{,}970.8tb(L + l)$

minimize $\quad f_2 := 136{,}432.2hlh^R + (68{,}216.1h^2 + 2{,}970.8tb)l^R$

minimize $\quad f_3 := 2{,}970.8tbL$

maximize $\quad f_4 := h^R$

maximize $\quad f_5 := l^R$

Constants:

$$\tau = \sqrt{\tau'^2 + \tau'\tau'' \tfrac{l}{R} + \tau''^2}, \quad \tau' = \tfrac{P}{\sqrt{2}hl}, \quad \tau'' = \tfrac{MR}{J}$$

$$M = P\left(L + \tfrac{l}{2}\right), \quad R = \sqrt{\tfrac{l^2 + (h+t)^2}{4}}, \quad J = \sqrt{2}hl\left(\tfrac{l^2}{6} + \tfrac{(h+b)^2}{2}\right)$$

$$\sigma = \tfrac{6PL}{t^2b}, \quad \delta = \tfrac{6PL^3}{Et^3b}, \quad P_c = \tfrac{4.013\sqrt{EI\alpha}}{L^2}\left(1 - \tfrac{t}{2L}\sqrt{\tfrac{EI}{\alpha}}\right), \quad I = \tfrac{tb^3}{12}, \quad \alpha = \tfrac{Gtb^3}{3}$$

$$E = 206.843(GPa), \quad G = 82.7371, \quad L^R = 0.01L(m)$$

Suppose that the ideal point f^* and the initial aspiration level $\overline{f}$ are given by

$$f^* = (2.08836,\ 0.3688,\ 0.01381,\ 0.00205,\ 0.07535),$$
$$\overline{f} = (2.2350,\ 0.450,\ 0.0150,\ 0.0010,\ 0.0600).$$

Then, the procedure for deciding the final solution according to the given aspiration level is as follows.

Stage 1. Generate Pareto optimal solutions by EMO based on the fitness evaluation of (ALGDEA_{fit}).

Stage 2. Pick up several Pareto optimal solutions from the set of Pareto optimal solutions obtained in Stage 1 as representative alternatives to the final decision making solution given by (6.1) and (6.2). In this example, representative alternatives are selected as follows:

$$A0 = (2.12170,\ 0.42431,\ 0.01413,\ 0.00134,\ 0.06519)$$
$$A1 = (2.09579,\ 0.41857,\ 0.01386,\ 0.00128,\ 0.06526)$$
$$A2 = (2.15548,\ 0.38580,\ 0.01436,\ 0.00108,\ 0.06076)$$
$$A3 = (2.09679,\ 0.41975,\ 0.01385,\ 0.00128,\ 0.06527)$$
$$A4 = (2.17270,\ 0.44531,\ 0.01441,\ 0.00152,\ 0.06623)$$
$$A5 = (2.20449,\ 0.44352,\ 0.01444,\ 0.00110,\ 0.07026)$$

Figure 6.13 represents the above alternatives normalized by

6.2 MCDM with Aspiration Level Method and GDEA

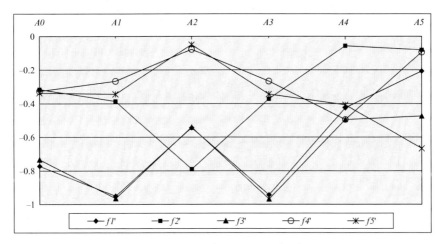

Fig. 6.13 Representative alternatives to the aspiration level

$$fi' := \frac{f_i - \overline{f}_i}{\overline{f}_i - f_i^*}, \quad i = 1, \ldots, 5.$$

These values mean the degree how much x does not attain to the aspiration level of decision maker. The fact that the values are all negative in this example shows that the aspiration level is mild as a whole, and thus the representative alternatives achieve the given aspiration level.

The alternative $A0$ is the Pareto optimal value closest to the aspiration level, and is regarded as the first candidate of decision making solution. However, if decision makers are not satisfied with the alternative $A0$, they search a more preferable alternative in a neighborhood of the alternative $A0$. Suppose that they want to improve some of objective functions at the alternative $A0$. Since Pareto optimal solutions cannot improve all objective functions simultaneously any more, decision makers have to agree with some sacrifice for some of other objective functions. How much some of objective functions should be improved and how much some of other objective functions should be sacrificed? This is the trade-off analysis. Seeing the alternatives $Ai, i = 1, \ldots, m$, it is possible to grasp a rough information of the trade-off and to know how much some objective function should be improved.

The procedure is terminated if decision makers are satisfied with one of them. Otherwise, proceed to the next stage.

Stage 3. This stage decides the direction to which decision makers want to move.

Suppose that decision makers want to improve f_2 at the alternative $A0$, because the degree of attainment for f_2 is the worst among those of all

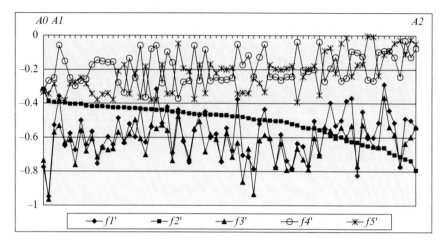

Fig. 6.14 Alternatives to decision maker's requests

objective functions. Then, among the representative alternatives obtained in Stage 1, f_2 of the alternatives *A1, A2* and *A3* are better than the alternative *A0*. Since the value of f_2 at *A3* is not so improved, it is preferable to move to a direction between *A1* and *A2*. Choose all Pareto optimal individuals P_i for which $f_2(P_i)$ is between $f_2(A1)$ and $f_2(A2)$. For those Pareto optimal individuals, Fig. 6.14 illustrates Pareto optimal values sorted in descending order of the value f_2.

It usually happens in multiple criteria decision making problems that improving an objective function makes other objective functions much worse than expected. Hence, proceed to the next stage on the basis of the trade-off relation among objective functions.

Stage 4. With the obtained result in Stage 3, ask decision makers' demands for other objective functions.

For example, suppose that decision makers want to keep f_4 and f_5 within some allowable range:

- The degree of attainment for f_4 should be less than -0.2.
- The degree of attainment for f_5 should be less than -0.25.

Then, pick up ones satisfying these requests of decision makers from the alternatives obtained in Stage 3. Figure 6.15 represents these alternatives in order of the value of f_2. Decision makers can finally choose a solution among them.

If they cannot decide their final solution, we go back to Stage 1 setting a new aspiration level.

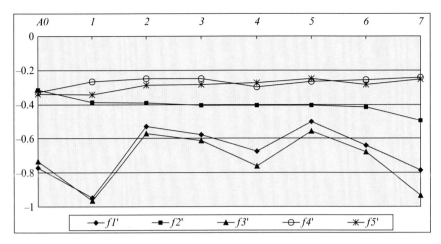

Fig. 6.15 Alternatives to decision maker's additional requests

6.3 Discussions

In this section, we introduced the method which utilizes aspiration level and generalized data envelopment analysis for supporting a multiple criteria decision making. Summarizing the feature of the method, we have the following:

1. The stated method can provide the most interesting part of Pareto optimal solutions close to the given aspiration level for decision makers.
2. It is possible to grasp the trade-off among objective functions in this area by selecting representative alternatives $A0$ and Ai, $i = 1, \ldots, r$, by (6.1) and (6.2).
3. The final solution can be decided on the basis of the provided information.

The aim of the stated method is to support decision making by showing the most interesting part of Pareto optimal solutions for decision makers. They can choose the final solution from the shown Pareto optimal solutions by making trade-off analysis on the visualized information even with many objective functions.

The basic idea of interactive multiobjective optimization methods is to search a decision solution by sequential interaction between decision makers' input of their aspiration levels and the output of a corresponding Pareto solution with some trade-off information from the DSS (decision support system). However, inputting several possible representative aspiration levels at a time in advance, one may expect the number of interactions would decrease. Indeed, in many cases we can get a better approximation of Pareto frontier in a larger range under inputs of multiple aspiration levels.

In many practical problems, however, decision makers know the outline of the problem through their experiences. In other words, decision makers have a rough estimate of a reasonable aspiration level in their mind. Moreover, a satisfactory solution can be usually obtained with only few iterations in many cases. In addition, it is time consuming to obtain a good approximation of Pareto frontier in a wide range.

Finally, we recommend that:

1. If a decision maker has a reasonable aspiration level in mind in advance, then the interactive iteration with a single aspiration level may be applied in sequence.

2. If decision makers want an innovative solution which cannot be anticipated in advance or have not a full knowledge on the problem (and hence on what their aspiration levels should be), plural representative aspiration levels may be adopted.

Chapter 7
Engineering Applications

This chapter points out several issues for industrial applications of multiobjective optimization along several examples through the authors' experience in collaborating with Japanese industrial companies. Some of major difficulties in engineering problems are (1) we have many (sometimes, a large number of) objective functions and (2) the function form of criteria is black box, namely cannot be explicitly given in terms of design variables.

In Sect. 2.3, we mentioned that the aspiration level approach to interactive multiobjective programming method is effective for the problem (1) above along an example of erection management problem of cable-stayed bridges. In particular, aspiration level methods using graphical user interface can be effectively used even in cases with a large number of objective functions.

Regarding the problem (2) above, sequential approximate optimization techniques can be effectively applied. Since criteria functions sometimes, in particular in engineering design, are highly nonlinear, they are predicted effectively by virtue of RBFN or SVR stated in the previous chapters.

In this chapter, we discuss how well the methodology stated in this book can be applied for practical engineering problems along examples of the reinforcement of cable-stayed bridges and the startup scheduling problems for thermal power plants.

7.1 Reinforcement of Cable-Stayed Bridges

After the big earthquake in Kobe in 1995, many in-service structures are required to improve their antiseismic property. The criterion in the *Specifications for Highway Bridges* for resisting huge earthquake called Level 2 Earthquake, which has little possibility to occur during the life of bridges, was revised drastically in Japan. According to the revision, relatively small and simple bridges in service, such as viaducts in city area, were verified their antiseismic property and almost of them have been reinforced. On the other hand, it is very difficult for large and/or complicated bridges, such as suspension bridge, cable-stayed bridge, arch bridge and so on, to be made a reinforcement because of impractical executing method and complicated dynamic response.

Recently many kinds of antiseismic device have been developed. It is practical for bridges to be installed a number of small devices taking into account of strength and/or space, and to obtain the most reasonable arrangement and capacity of the devices by using optimization techniques. In this problem, the form of objective function is not given explicitly in terms of design variables, but the value of the function is obtained by seismic response analysis. Since this analysis needs much cost and long time, it is strongly desirable to make the number of analysis as few as possible.

To this end, the described method in the preceding sections is applied. In this example, radial basis function networks (RBFN) are employed in predicting the form of objective function, and genetic algorithms (GA) in searching the optimal value of the predicted objective function. The method of Sect. 5.6 is applied to a problem of antiseismic improvement of a cable-stayed bridge which typifies the difficulty of reinforcement of in-service structure [109]. In this investigation, we determine an efficient arrangement and amount of additional mass for cables to reduce the seismic response of the tower of a cable-stayed bridge (Fig. 7.1).

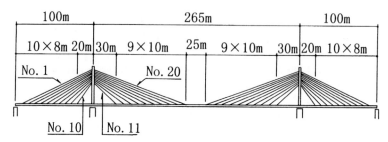

Fig. 7.1 Cable-stayed bridge

7.1.1 Dynamic Characteristics of Cable-Stayed Bridge

Though modal coupling among structural elements of cable-stayed bridge, i.e., girder, towers and cables, makes its seismic response complicated, this fact shows that the seismic response can be controlled by dynamic characteristics of each element. Obviously it is the dynamic characteristic of cable that we can change more easily. There are two ways to change characteristics of cables, i.e., adding mass to cable and tensing or loosening cable. The former can be acceptable if the amount of mass is limited appropriately [129],

7.1 Reinforcement of Cable-Stayed Bridges

but the latter cannot be alternative because it causes bridge an undesirable stress condition. Figure 7.2 shows an impression of additional mass with cable.

Fig. 7.2 Cable with additional mass

The influence of additional mass on cables is investigated by numerical sensitivity analysis. Analytical model shown in Fig. 7.1 is three-span continuous and symmetrical cable-stayed bridge whose 2 × 20 cables are in one plane and the towers stand freely in their transverse direction. The mass must be distributed over cables uniformly to prevent them from concentrating deformation.

Seismic response to be paid attention is the stress at the fixed end of the tower when earthquake occurred in transverse direction. Seismic response analysis is carried out by spectrum method. As there are a lot of modes whose natural frequencies are close to each other, response is evaluated by the complete quadratic combination method. Input spectrum is given in the new Specifications for Highway Bridges in Japan.

The natural frequencies of modes accompanied with bending of tower (natural frequency of tower alone is 1.4 Hz) range from 0.79 to 2.31 Hz due to coupling with cables. For example, Fig. 7.3 demonstrates half part of symmetrical mode shapes which are coupling among tower, girder and cables.

Change of bending moment at the fixed end of tower with additional mass on one cable distributed uniformly on it is shown in Fig. 7.4. The abscissa means the ratio of additional mass to cable's own mass ranged from 0.0 to 1.0, and the ordinate means response of bending moment at the fixed end of tower. The second upper cable (No. 19) in the center span has much more influence on it than any other cables. Then, change of seismic response of bending moment of interest is considered individually with additional mass on each cable when additional mass ratio of No. 19 is fixed to 0.5. The result is shown in Fig. 7.5. No. 20 cable becomes effective instead of No. 1 and No. 2. The minimum value of bending moment is 64.6 MN·m which is 44% of bending moment without additional mass. As mentioned above, the seismic response of tower can be controlled by additional mass to cables, but each cable influences to one another complexly. Then, the reasonable distribution of additional mass must be decided by optimization technique.

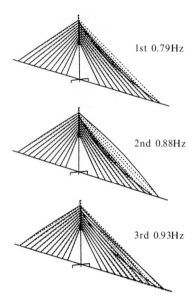

Fig. 7.3 Examples of mode shape

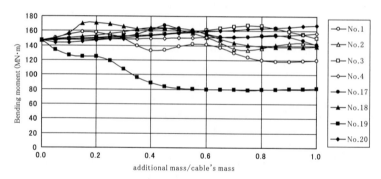

Fig. 7.4 Influence on bending moment by additional mass (no ratio of additional mass is fixed)

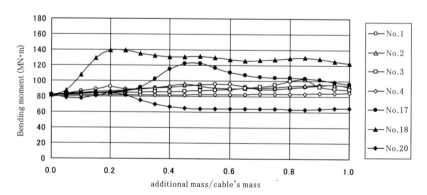

Fig. 7.5 Influence on bending moment by additional mass (No. 19 ratio of additional mass is fixed at 0.5)

7.1 Reinforcement of Cable-Stayed Bridges

Case 1

The objective is to minimize the bending moment M at the base of tower. The variables are ratios of additional mass and mass of cables. The number of variables is 20. The lower bound and upper bound of each variable are 0.0, and 1.0, respectively. For comparison, we apply a quasi-Newton method based on numerical differentials as an existing method in five trials with different initial points in order to obtain a global optimum.

We use RBFN for prediction, and GA for optimization of predicted objective function in SAO. As a genetic algorithm, in particular, BLX-α is adopted, because it is observed to be effective for continuous variables. The population is 10, and the number of generation is 200. We set $\lambda = 0.01$, and decided the value of width r of Gaussian basis function by the simple estimate given by (7.1):

$$r = \frac{d_{max}}{\sqrt[n]{nm}},\qquad(7.1)$$

where d_{max} is the maximal distance among the data; n is the dimension of data; m is the number of data (see Sect. 4.2 for details).

Starting the iteration with 60 sample points, the first 20 sample points are generated randomly with one of variables fixed at the upper bound 1 by turns; the next 20s are generated similarly with one of variables fixed at the lower bound 0 by turns; the last 20s similarly with one of variables fixed at the midvalue 0.5 by turns. The parameters for convergence are $C_x^0 = 20$, $C_f^0 = 20$ and $l_0 = 0.1$.

The result is shown in Table 7.1. It is seen that the method [102, 103] by SAO can find out fairly good solutions within 1/10 or less times of analysis than the conventional optimization.

Case 2

Now, we take the number of cables to be added with masses, N, as another objective function in addition to the bending moment M. Namely, the objective function is

$$F = \frac{M}{M_0} + \alpha \left(\frac{N}{N_0}\right),$$

where α is a parameter for trade-off between the first term and the second one. M_0 and N_0 are used for normalization of the bending moment and the number of cables, respectively. In this experiment, setting $M_0 = 147.0$MN·m and $N_0 = 20$, a simple GA is used, because some of variables are discrete. The parameters for calculation are the same as in Case 1. The result is given in

Table 7.1 Result for Case 1

		Existing method	RBF network	
			Best	Average
	1	0.32	0.04	0.40
	2	1.00	0.69	0.84
	3	0.49	0.18	0.51
	4	0.62	0.82	0.80
	5	0.81	0.57	0.64
	6	0.52	0.43	0.56
	7	0.49	1.00	0.39
	8	0.52	0.44	0.66
	9	0.48	0.94	0.50
Cable	10	0.48	0.50	0.56
No.	11	0.50	0.45	0.47
	12	0.55	1.00	0.74
	13	0.70	0.85	0.71
	14	0.61	0.50	0.30
	15	0.61	1.00	0.58
	16	0.46	0.24	0.37
	17	0.22	0.10	0.13
	18	1.00	0.95	0.91
	19	0.98	1.00	0.94
	20	1.00	1.00	0.91
Bending moment (MN·m)		50.3	54.90	63.70
No. of analysis		1,365	150.00	124.80

Table 7.2. It should be noted that the number of analysis in the method [102, 103] using SAO is reduced to about 1/20 of the conventional method. Although the precision of solution by SAO is behind the conventional method, it is sufficiently acceptable in practice.

7.1.2 Discussions

In cable-stayed bridge as an example of structure in service, effective countermeasure against earthquake was investigated by a computational intelligence. The conclusions can be summarized as follows:

1. Additional mass to cables can sharply reduce bending moment at the fixed end of tower when earthquake occurs in transverse direction.

7.1 Reinforcement of Cable-Stayed Bridges

Table 7.2 Result for Case 2

		Existing method		RBF network	
		Best	Average	Best	Average
	1	0.00	0.00	0.00	0.83
	2	0.00	0.00	0.00	0.09
	3	0.00	0.00	0.00	0.00
	4	0.00	0.00	0.00	0.04
	5	0.00	0.00	0.00	0.00
	6	0.00	0.00	0.00	0.00
	7	0.00	0.00	0.00	0.00
	8	0.00	0.00	1.00	0.99
	9	0.00	0.00	0.00	0.10
Cable	10	0.00	0.00	0.86	0.53
No.	11	0.00	0.00	0.00	0.00
	12	0.00	0.00	1.00	0.63
	13	0.00	0.00	0.00	0.13
	14	0.00	0.00	0.86	0.53
	15	0.00	0.00	0.00	0.00
	16	0.00	0.00	0.00	0.00
	17	0.00	0.00	0.00	0.00
	18	0.00	0.00	0.00	0.00
	19	0.71	0.74	1.00	1.00
	20	0.86	0.83	0.86	0.79
Bending moment (MN·m)		62.6	62.8	67.1	69.7
No. of cable with additional mass		2	2	6	6.2
Objective function		0.526	0.527	0.756	0.784
No. of analysis		4,100	3,780	199	193.3

2. As there are a lot of conditions of distribution of additional mass, the procedure which reasonably decided distribution by optimization technique must be considered.

3. Application of conventional optimization technique compells us to analyze seismic response very many times, because objective function is not given explicitly in terms of design variables, but obtained by some analysis.

4. In the problem of cable-stayed bridge, the method of sequential approximate optimization [102, 103] can find out fairly good solutions within $1/10 \sim 1/20$ or less times of analysis than the conventional methods.

7.2 Multiobjective Startup Scheduling of Power Plants

Typically, the *startup scheduling problem* of thermal power plants has several
conflicting objectives, such as startup time, fuel consumption rate, lifetime
consumption rate of machines, and pollutant emissions rate. These criteria
are affected by the varying market price of electricity as well as fuel, main-
tenance, and environmental costs. Therefore, it is important to achieve a
flexible startup schedule based on multiple criteria decision making in the
overall plant management strategy.

The startup characteristics are evaluated by using some dynamic sim-
ulators; however, determining the optimal startup schedule is complicated
because it is necessary to iterate the dynamic simulation on the basis of trial
and error using the engineer's intuition and experience. Several methods for
optimizing the startup schedule have been proposed. For instance, a fuzzy
expert system [88], a genetic algorithm with enforcement operation [67], and
a nonlinear programming technique [138]. However, all these works aimed to
optimize a single-objective function (e.g., only the startup time is minimized
under the operational constraints).

On the other hand, Shirakawa et al. [140] proposed to apply a combination
of SAO and aspiration level method to cases with multiobjective functions.
In the following, we introduce their optimal startup scheduling system for a
thermal power plant. The system consists of dynamic simulation, metamodeling, and an interactive multiobjective programming technique.

7.2.1 Startup Scheduling for Combined Cycle Power Plant

Consider a multishaft type combined cycle power plant, as shown in Fig. 7.6.
It consists of three gas turbine units, three heat recovery steam generator
(HRSG) units, and one steam turbine unit. The gas turbines and the steam
turbine drive the generators. Also, the HRSGs generate steam for the steam
turbine using waste heat from the gas turbines. This plant generates a total
output of 670 MW.

An example of startup schedule is shown in Fig. 7.7. The decision variables
are:

1. Steam turbine acceleration rate (x_1)
2. Low-speed heat soak time (x_2)
3. High-speed heat soak time (x_3)
4. Initial-load heat soak time (x_4)

7.2 Multiobjective Startup Scheduling of Power Plants

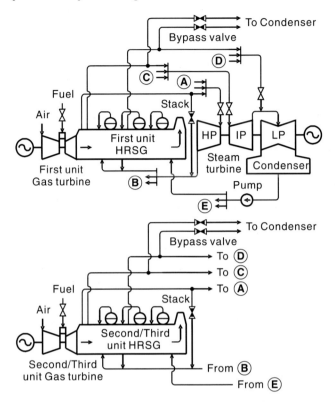

Fig. 7.6 Plant configuration

where the steam turbine acceleration rate is discrete among {120, 180, 360} rpm/min, while each heat soak time is continuous between 50 and 60 min.

The objective functions are:

1. Startup time (f_1)
2. Fuel consumption rate (f_2)
3. Thermal stress of the steam turbine rotors (f_3)

Our aim is to minimize the startup time and the fuel consumption rate. In addition, the less the thermal stress of the steam turbine rotors is, the longer the service lifetime of the stream turbine is. However, these objective functions have a trade-off depending on the above-mentioned schedule variables x_1 to x_4. As the acceleration rate x_1 increases and the heat soak times x_2, x_3 and x_4 decrease, both the startup time and fuel consumption rate decrease; however, the thermal stress of the steam turbine rotors increases. This trend of the fuel consumption rate becomes more significant as the initial-load heat soak time x_4 is varied because the gas turbine load is higher with x_4 than with the other schedule variables x_1, x_2 and x_3.

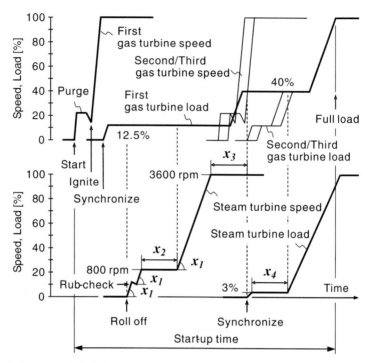

Fig. 7.7 Startup schedule for a power plant

There exist many operational constraints in this plant. However, most of those operational constraints are safely controlled within the limits at any time (e.g., the drum water level and the steam temperature). As a result, the operational constraints can consider only the thermal stress of the steam turbine rotors and the NO_x emission rate from the plant. The thermal stress of the steam turbine rotors in the above-mentioned objective functions has an upper limit to prevent metal creep and fatigue. The NO_x emission rate from the plant increases significantly with the rapid startup; however, it has an upper limit in accordance with the environmental regulations.

7.2.2 Sequential Approximate Multiobjective Optimization for Optimal Startup Scheduling

Criteria such as objectives and constraints are evaluated via some simulator based on dynamic models derived from thermohydraulic conservation equations. Shirakawa et al. [139, 140] developed a high accuracy plant simulator using MATLAB TM/Simulink TM. However, the high accuracy plant

7.2 Multiobjective Startup Scheduling of Power Plants

simulator is extremely time consuming because detailed, large-scale, and nonlinear models are used. Therefore, sequential approximate multiobjective optimization techniques are expected to be appropriate in order to decrease the number of simulations as much as possible.

The algorithm consists of the following steps:

Step 1. The training data for RBFN are prepared by dynamic simulation. These training data consist of the values of schedule variables and criteria functions.

Step 2. The user sets the aspiration level of the objective functions.

Step 3. RBFN is trained on the basis of the data set obtained at Step 1 to create approximate functions for the criteria. The approximate function of the augmented Tchebyshev scalarization function F given by (2.2) (Sect. 2.3.1) to the given aspiration level is created by either of the following methods:

1. The method that directly acquires the form of F by using RBFN

2. The method that calculates F using created approximate criteria functions

Step 4. The startup schedule closest to the aspiration level is searched by solving the auxiliary scalar optimization (AP) (explained in Sect. 2.3.1) with the predicted approximate functions by using GA: This process yields an approximate Pareto optimal solution closest to the given aspiration level.

Step 5. In order to evaluate the approximation errors, the dynamic simulation is executed according to the obtained startup schedule. If the approximation errors are large, some additional training points for RBFN are provided in the neighborhood of the obtained startup schedule. Further, Steps 1–5 are repeated until the approximation errors become small. If the approximation errors are sufficiently small, the obtained startup schedule is displayed to the user.

Step 6. The user judges whether or not the obtained startup schedule is satisfactory. If the user is not satisfied, the user modifies the aspiration level. Further, Steps 2–6 are repeated until the user obtains an agreeable startup schedule. If the user is satisfied with the present solution, then the iteration is terminated.

7.2.3 Application Results

Experimental studies have been executed for a warm startup condition, i.e., the initial temperature of the steam turbine rotors is 180–290°C. The steam turbine acceleration rate x_1, low-speed heat soak time x_2, high-speed heat soak time x_3, and initial-load heat soak time x_4, are treated as schedule variables (see Fig. 7.7). The startup time f_1 is from the start of the first gas turbine to the plant base load operation. The fuel consumption rate f_2 is the gross weight during startup. The thermal stress of the steam turbine rotors, $f_3(=g_{c1})$, is the maximum value during startup. The NO_x emission rate from the plant g_{c2} is the maximum value of moving average per hour during startup. Further, g_{c1} and g_{c2} have upper limits g_{U1} and g_{U2}, respectively. The problem is to find the vector of the schedule variables $\boldsymbol{x} = (x_1,\ x_2,\ x_3,\ x_4)^T$, in which the objective functions f_1, f_2 and f_3, are to be minimized under the operational constraints $g_{c1} \leq g_{U1}$ and $g_{c2} \leq g_{U2}$. This problem is summarized as follows:

$$\underset{\boldsymbol{x}}{\text{minimize}} \qquad \boldsymbol{f}(\boldsymbol{x}) := (f_1(\boldsymbol{x}),\ f_2(\boldsymbol{x}),\ f_3(\boldsymbol{x}))^T$$
$$\text{subject to} \qquad g_1(\boldsymbol{x}) = g_{c1}(\boldsymbol{x}) - g_{U1} \leqq 0,$$
$$g_2(\boldsymbol{x}) = g_{c2}(\boldsymbol{x}) - g_{U2} \leqq 0,$$

where

$$x_1 \in \{20,\ 180,\ 360\}\ \text{rpm/min},$$
$$5\ \text{min} \leqq x_2 \leqq 60\ \text{min},$$
$$0\ \text{min} \leqq x_3,\ x_4 \leqq 60\ \text{min}.$$

Training of RBFN

The dynamic simulation executes 250 startup schedules randomly in order to prepare data sets of schedule variables and criteria functions. The data sets were divided into two parts. The first (80% of the data sets) was used as training data for RBFN, and the second (20% of the data sets) as a validation test. RBFN was trained for all criteria f_1, f_2, f_3, g_{c1} and g_{c2} based on the training data. The validation of approximation by RBFN is checked by Fig. 7.8. Figure 7.8a–c shows the relation of the predicted values and actual ones for f_1, f_2 and f_3, respectively, made dimensionless by the upper limit. Figure 7.8d shows the one for the augmented Tchebyshev scalarization function F of (2.2) with the aspiration level $(\overline{f}_1,\ \overline{f}_2,\ \overline{f}_3) = (200\,\text{min},\ 74.4\,\text{ton},\ 94.9\%)$. The average and maximum values of the approximation error defined by the relative error are summarized in Table 7.3. The approximate functions created from RBFN agree well with the actual dynamic simulation results.

Multiobjective Optimization

Let the first aspiration level be $(\overline{f}_{1,1},\ \overline{f}_{2,1},\ \overline{f}_{3,1}) = (200\,\text{min},\ 74.4\,\text{ton},\ 94.9\%)$. Then, the first solution is $(f_{1,1},\ f_{2,1},\ f_{3,1}) = (175\,\text{min},\ 71.6\,\text{ton},\ 88.5\%)$.

7.2 Multiobjective Startup Scheduling of Power Plants

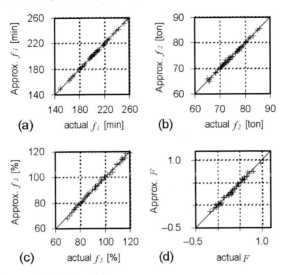

Fig. 7.8 Validation of RBFN model

Table 7.3 Approximation error of the RBFN models (unit: %)

Error	f_1	f_2	f_3, g_{c1}	g_{c2}
Average	0.15	0.37	0.66	0.16
Maximum	1.13	1.93	2.68	0.56

Here, the symbols □ and ■ in Fig. 7.9 show the given aspiration level and the resulting solution, respectively.

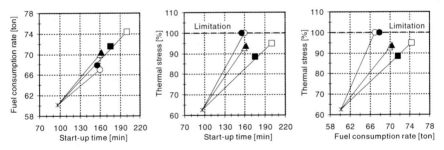

Fig. 7.9 Aspiration levels and corresponding solutions

Now, suppose that the user wants to further decrease the value of f_1 significantly and that of f_2 slightly. Since the present solution is already one of the Pareto solutions, it is impossible to improve all the criteria. Therefore, suppose that the user agrees to relax f_3 and let the second aspiration level be $(\overline{f}_{1,2}, \overline{f}_{2,2}, \overline{f}_{3,2}) = (160\,\text{min}, 70.0\,\text{ton}, 92.8\%)$. Then, the second solution to the revised aspiration level is $(f_{1,2}, f_{2,2}, f_{3,2}) = (162\,\text{min}, 70.3\,\text{ton}, 93.6\%)$.

Here, the symbols ○ and ● in Fig. 7.9 show the given aspiration level and the resulting solution, respectively. Although the obtained solution does not completely attain the aspiration level of f_1 and f_2, it should be noted that the solution is improved than the previous one. The reason why the improvement of f_1 and f_2 does not agree with the wish of the user seems because the amount of relaxation of f_3 is not large enough to compensate the improvement of f_1 and f_2.

Furthermore, therefore, suppose the third aspiration $(\overline{f}_{1,3}, \overline{f}_{2,3}, \overline{f}_{3,3}) = (160\,\text{min}, 67.0\,\text{ton}, 100.0\%)$. Then the resulting solution is $(f_{1,3}, f_{2,3}, f_{3,3}) = (156\,\text{min}, 67.9\,\text{ton}, 100.0\%)$. Here, the symbols △ and ▲ in Fig. 7.9 show the given aspiration level and the resulting solution, respectively. The improvement in f_2 does not match the requirement of the user because f_3 reaches the upper limit. This solution is an operating limitation of this plant, where both the startup time and fuel consumption rate cannot be reduced any further.

The optimal startup schedule of the second solution is demonstrated in the left-hand side of Fig. 7.10, while the right-hand side shows the result of the actual dynamic simulation for $(x_1, x_2, x_3, x_4) = (180\,\text{rpm/min}, 40.0\,\text{min}, 40.0\,\text{min}, 30.0\,\text{min})$. In this figure, GT denotes the gas turbines; ST, the steam turbine; Stress, the normalized thermal stress of the steam turbine rotors; and NOx, the normalized NOx emission rate from the plant. The results are summarized in Table 7.4, where Solution (RBFN) represents the results of the RBFN model, and Solution (Actual) represents the results of the actual dynamic simulation. Both the first solution and second solution are the Pareto solutions that are startup schedules closest to each aspiration level and satisfy the operational constraints.

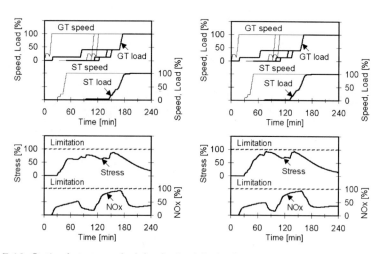

Fig. 7.10 Optimal startup schedule obtained for each aspiration level

7.2 Multiobjective Startup Scheduling of Power Plants

Table 7.4 Results of the objective functions

	f_1 (min)	f_2 (ton)	f_3 (%)
Aspiration level No. 1	205	74.0	95.0
Solution (RBFN) No. 1	186	71.8	88.4
Solution (Actual) No. 1	186	71.5	87.3
Aspiration level No. 2	165	69.0	92.0
Solution (RBFN) No. 2	167	69.4	93.7
Solution (Actual) No. 2	166	69.3	92.8

7.2.4 Discussions

It has been observed that the stated interactive decision support system is effective for the optimal startup scheduling for thermal power plants in view of the following two points:

1. Since the high accuracy simulator needs a considerable computation time, it is almost unrealistic to apply only it for getting an optimal solution. The stated system can provide an approximate model (metamodel) in a satisfactory precision with reasonably less number of simulations.

2. Aspiration levels are intuitive for the decision maker to answer. Moreover, since the obtained solution is nearest to the given aspiration level, the decision maker can understand the structure of the feasible region through the iteration. Due to this learning effect, the final solution can be obtained in a small number of iterations in most cases.

This feature of less simulations and less interactions makes the stated method to be effective for optimal startup scheduling in power plant, although it is not limited in the case but applicable to a wide range of practical problems.

References

1. Aitken, A C (1934) The normal form of compound and induced matrices. In: Proceedings of the London Mathematical Society II, pp. 354–376
2. Akaike, H (1974) A new look at statistical model identification. IEEE Transactions on Automatic Control 19:716–723
3. Arakawa, M, Hagiwara, I and Yamakawa, H (1998) Robust design using fuzzy numbers. In: Proceedings of DETC'98-DAC14536 (in CD-ROM)
4. Arakawa, M, Nakayama, H, Hagiwara, I and Yamakawa, H (1998) Multiobjective optimization using adaptive range genetic algorithms with data envelopment analysis. In: A Collection of Technical Papers on 7th Symposium on Multidisciplinary Analysis and Optimization (TP98-4970), AIAA, vol. 3, pp. 2074–2082
5. Asada, T and Nakayama, H (2003) SVM using multi-objective linear programming and goal programming. In: Tanino, T, Tanaka, T and Inuiguchi, M (eds) Multi-Objective Programming and Goal Programming. Springer, Berlin, pp. 93–98
6. Banker, R D, Charnes, A and Cooper, W W (1984) Some models for estimating technical and scale inefficiencies in data envelopment analysis. Management Science 30:1078–1092
7. Benayoun, R, de Montgolfier, J, Tergny, J and Larichev, O (1971) Linear programming with multiple objective functions: STEP method (STEM). Mathematical Programming 1(3):366–375
8. Bennett, K P and Mangasarian, O L (1992) Robust linear programming discrimination of two linearly inseparable sets. Optimization Methods and Software 1:23–34
9. Betrò, B (1991) Bayesian methods in global optimization. Journal of Global Optimization 1(1):1–14
10. Bishop, C M (2006) Pattern Recognition and Machine Learning. Springer, Berlin
11. Borde, J and Crouzeix, J P (1987) Convergence of a Dinkelbach-type algorithm in generalized fractional programming. Zeitschrift fur Operations Research 31:A31–A54
12. Box, G E P and Draper, N R (2007) Response Surfaces, Mixtures, and Ridge Analysis, 2nd edn. Wiley, New York
13. Branke, J, Deb, K, Miettinen, K and Slowinski, R (2008) Multiobjective Optimization Interactive and Evolutionary Approaches. Lecture Notes in Computer Science. Springer, Berlin
14. Breiman, L, Friedman, J, Olsshen, C R and Stone, C (1984) Classification and Regression Trees. Chapman and Hall/CRC, Boca Raton, FL
15. Buhmann, M D (2003) Radial Basis Functions. Cambridge University Press, Cambridge
16. Cavalier, T M, Ignizio, J P and Soyster, A L (1989) Discriminant analysis via mathematical programming: On certain problems and their causes. Computers and Operations Research 16(4):353–362

186 References

17. Charnes, A and Cooper, W (1961) Management Models and Industrial Applications of Linear Programming. Wiley, New York
18. Charnes, A, Cooper, W W and Rhodes, E (1978) Measuring the efficiency of decision making units. European Journal of Operational Research 2(6):429–444
19. Charnes, A, Cooper, W W and Rhodes, E (1979) Measuring the efficiency of decision making units. European Journal of Operational Research 3(4):339
20. Chen, W, Wiecek, M M and Zhang, J (1998) Quality utility – A compromise programming approach to robust design. In: Proceedings of DETC'98-DAC5601 (in CD-ROM)
21. Chen, V C P, Tsui, K L, Barton, R R and Allen, J K (2003) A review of design modeling in computer experiments. In: Handbook of Statistics 22: Statistics in Industry, pp. 231–261
22. Cherkassky, V and Mulier, F (1998) Learning from Data: Concepts, Theory, and Methods. Adaptive and Learning Systems for Signal Processing, Communications and Control Series. Wiley-Interscience, New York
23. Coello, C A C (1999) An updated survey of evolutionary multiobjective optimization techniques: State of the art and future trends. In: Proceedings of Congress on Evolutionary Computation (CEC-1999), pp. 3–13
24. Coello, C A C, Van Veldhuizen, D A and Lamont, G B (2001) Evolutionary Algorithms for Solving Multi-Objective Problems. Kluwer, Boston, MA
25. Cortes, C and Vapnik, V (1995) Support vector networks. Machine Learning 20(3):273–297
26. Cressie, N A G (1993) Statistics for Spatial Data. Wiley, New York
27. Cristianini, N and Shawe-Taylor, J (2000) An introduction to support vector machines: And other kernel-based learning methods. Cambridge University Press, Cambridge
28. Dantzig, G B (1983) Reminiscences about the origins of linear programming. In: Bachen, A, Grotschel, M and Korte, B (eds) Mathematical Programming – The State of the Art. Springer, Berlin, pp. 247–263
29. De Jong, K A (1975) An Analysis of the Behavior of a Class of Genetic Adaptive Systems. Ph.D. Thesis, University of Michigan
30. Deb, K (2001) Multi-Objective Optimization Using Evolutionary Algorithms. Wiley, New York
31. Deb, K, Agrawal, S, Pratap, A and Meyarivan, T (2000) A fast elitist non-dominated sorting genetic algorithm for multi-objective optimization: NSGA-II. In: Proceedings of Parallel Problem Solving from Nature VI (PPSN-VI), pp. 849–858
32. Deb, K, Pratap, A, Agarwal, S and Meyarivan, T (2000) A Fast and Elitist Multi-Objective Genetic Algorithm: NSGA-II. Technical Report 200001, Indian Institute of Technology, Kanpur Genetic Algorithms Laboratory (KanGAL), Kanpur
33. Deb, K, Pratap, A, Agarwal, S and Meyarivan, T (2002) A fast and elitist multiobjective genetic algorithm: NSGA-II. IEEE Transactions on Evolutionary Computation 6(2):182–197
34. Deprins, D, Simar, L and Tulkens, H (1984) Measuring labor-efficiency in post offices. In: Marchand, M, Pestieau, P and Tulkens, H (eds) The Performance of Public Enterprises: Concepts and Measurements. Elsevier Science, Amsterdam, pp. 247–263
35. Dixon, L C W and Szegö, G P (1978) The global optimization problem: An introduction. In: Towards Global Optimisation 2. North-Holland, Amsterdam, pp. 1–15
36. Edgeworth, F Y (1881) Mathematical Psychics: An Essay on the Application of Mathematics to the Moral Sciences. Kegan Paul, London
37. Ehrgott, M and Gandibleux, X (2002) Multiple Criteria Optimization – State of the Art Annotated Bibliographic Surveys. Kluwer, Dordrecht
38. Erenguc, S S and Koehler, G J (1990) Survey of Mathematical Programming Models and Experimental Results for Linear Discriminant Analysis. Managerial and Decision Economics 11:215–225

References 187

39. Eschenauer, H A, Koski, J and Osyczka, A (1990) Multicriteria design optimization. Springer, Berlin
40. Farrell, M J (1957) The measurement of productive efficiency. Journal of the Royal Statistical Society A 120:253–281
41. Fedorov, V V and Hackl, P (1997) Model-Oriented Design of Experiments. Lecture Notes in Statistics, Springer, Berlin
42. Ferland, A and Potvin, J (1985) Generalized fractional programming: Algorithms and numerical experimentation. European Journal of Operational Research 20(1):92–101
43. Fiacco, A V (1983) Introduction to Sensitivity Analysis in Nonlinear Programming. Academic, New York
44. Fonseca, C M and Fleming, P J (1993) Genetic algorithm for multiobjective optimization, formulation, discussion and generalization. In: Proceedings of the 5th International Conference: Genetic Algorithms, pp. 416–423
45. Freed, N and Glover, F (1981) Simple but powerful goal programming models for discriminant problems. European Journal of Operational Research 7(1):44–60
46. Furukawa, K, Inoue, K, Nakayama, H and Ishido, K (1986) Management of erection for cable stayed bridge using satisficing trade-off method. Journal of the Japan Society of Civil Engineers 374/I-6:495–502 (in Japanese)
47. Gal, T, Stewart, T J and Hanne, T (1999) Multicriteria Decision Making – Advances in MCDM Models, Algorithms, Theory, and Applications. Kluwer, Boston, MA
48. Gazut, S, Martinez, J M, Dreyfus, G and Oussar, Y (2008) Towards the optimal design of numerical experiments. IEEE Transactions on Neural Networks 19
49. Goldberg, D E (1989) Genetic Algorithms in Search, Optimization and Machine Learning. Addison-Wesley, Boston, MA
50. Goldberg, D E and Deb, K (1991) A comparative analysis of selection schemes used in genetic algorithms. In: Proceedings of Foundations of Genetic Algorithms 1 (FOGA-1), pp. 69–93
51. Grauer, M, Lewandowski, A and Wierzbicki, A P (1984) DIDASS theory, implementation and experiences. In: Grauer, M, Lewandowski, A and Wierzbicki, A P (eds) Proceedings of an International Workshop on Interactive Decision Analysis and Interpretative Computer Intelligence. Springer, Berlin, pp. 22–30
52. Haimes, Y Y, Hall, W A and Freedman, H B (1975) Multiobjective Optimization in Water Resources Systems, The Surrogate Worth Trade-off Method. Elsevier Scientific, Amsterdam
53. Hastie, T, Tibshirani, R and Friedman, J H (2001) The Elements of Statistical Learning: Data Mining, Inference, and Prediction. Springer Series in Statistics, Springer, Berlin
54. Haykin, S (1994) Neural Networks: A Comprehensive Foundation. Macmillan, New York
55. Haykin, S (1999) Neural Networks: A Comprehensive Foundation, 2nd edn. Prentice Hall, Upper Saddle River, NJ
56. He, S, Prempain, E and Wu, Q (2004) An Improved Particle Swarm Optimizer for Mechanical Design Optimization Problems. Engineering Optimization, 36(5): 585–605
57. Holland, J H (1975) Adaptation in Natural and Artificial Systems. University of Michigan Press, Ann Arbor, MI
58. Holland, J H (1992) Adaptation in Natural and Artificial Systems: An Introductory Analysis with Applications to Biology, Control, and Artificial Intelligence (Reprint Edition). MIT, Cambridge
59. Horn, J, Nafpliotis, N and Goldberg, D E (1994) A niched Pareto genetic algorithm for multiobjective optimization. In: Proceedings of the 1st IEEE Conference on Evolutionary Computation, IEEE World Congress on Computational Intelligence, vol. 1, pp. 82–87

60. Hsu, Y H, Sun, T L and Leu, L H (1995) A two-stage sequential approximation method for nonlinear discrete variable optimization. In: Proceedings of ASME Design Technical Conferences, pp. 197–202
61. Jahn, J (1996) Introduction to the Theory of Nonlinear Optimization. Springer, Berlin
62. Jin, Y (2005) A comprehensive survey of fitness approximation in evolutionary computation, Soft Computing 9(1):3–12
63. Johnson, M E, Moore, L M and Ylvisaker, D (1990) Minimax and maximin distance designs. Journal of Statistical Inference and Planning 26(1):131–148
64. Jones, D R (2001) A taxonomy of global optimization methods based on response surfaces. Journal of Global Optimization 19:345–383
65. Jones, D R, Schonlau, M and Welch, W J (1998) Efficient global optimization of expensive black-box functions. Journal of Global Optimization 13:455–492
66. Journel, A G and Huijbregts, C J (1978) Mining Geostatistics. Academic, New York
67. Kamiya, A, Kawai, K, Ono, I and Kobayashi, S (1999) Adaptive-edge search for power plant start-up scheduling. IEEE Transactions on System, Man, and Cybernetics, Part C: Applications and Reviews 29(4):518–530
68. Kannan, B K and Kramer, S N (1994) An augmented Lagrange multiplier based method for mixed integer discrete continuous optimization and its applications to mechanical design. ASME Transactions, Journal of Mechanical Design 116(2):405–411
69. Kitayama, S, Arakawa, M and Yamazaki, K (2006) Penalty function approach for the mixed discrete non-linear problems by particle swarm optimization. Structural and Multidisciplinary Optimization 32(3):191–202
70. Kleijnen, J P C (2007) Design and Analysis of Simulation Experiments. International Series in Operations Research & Management Science, Springer, Berlin
71. Knowles, J and Nakayama, H (1984) Meta-modeling in multi-objective optimization. In: Branke, J, Deb, K, Miettinen, K and Slowinski, R (eds) Multiobjective Optimization Interactive and Evolutionary Approaches. Springer, Berlin
72. Knowles, J D and Corne, D W (1999) The Pareto archived evolution strategy: A new baseline algorithm for Pareto multiobjective optimisation. In: Proceedings of Congress on Evolutionary Computation (CEC-1999), pp. 98–105
73. Konno, H and Inori, M (1987) Bond Portfolio Optimization by Bilinear Fractional Programming. Technical Report, Research Paper IKSS 87-4, Institute of Human and Social Sciences, Tokyo Institute of Technology
74. Koopmans, T C (1951) Analysis of production as an efficient combination of activities. In: Koopmans (ed) Activity Analysis of Production and Allocation. Wiley, New York, pp. 33–97
75. Korhonen, P and Wallenius, J (1988) A Pareto race. Naval Research Logistics 35(6):615–623
76. Krige, D G (1951) A statistical approach to some mine valuations and allied problems at the Witwatersrand, Master's thesis of the University of Witwatersrand
77. Kuhn, H W and Tucker, A W (1951) Nonlinear programming. In: Proceedings of the 2nd Berkeley Symposium on Mathematical Statistics and Probability, pp. 481–492
78. Kushner, H J (1964) A new method of locating the maximum of an arbitrary multi-peak curve in the presence of noise. Journal of Basic Engineering 86(1):97–106
79. Lasdon, L S (1970) Optimization Theory for Large Scale. Macmillan, New York
80. Lin, S S, Zhang, C and Wang, H P (1995) On mixed-discrete nonlinear optimization problems: A comparative study, Engineering Optimization 23(4):287–300
81. Locatelli, K (1997) Bayesian algorithms for one-dimensional global optimization. Journal of Global Optimization 10(1):57–76
82. Mackey, M C and Glass, L (1977) Robust linear programming discrimination of two linearly inseparable sets. Science 197:287–289
83. Mangasarian, O L (1968) Multisurface method of pattern separation. IEEE Transactions on Information Theory IT-14:801–807
84. Mangasarian, O L (1999) Arbitrary-norm separating plane. Operations Research Letters 24:15–23

References 189

85. Marcotte, P and Savard, G (1992) Novel approaches to the discrimination problem. ZOR – Methods and Models of Operations Research 36:517–545
86. Martin, J D and Simpson, T W (2005) Use of Kriging models to approximate deterministic computer models. AIAA Journal 43(4):853–863
87. Matheron, G (1963) Principles of geostatistics. Economic Geology 58:1246–1266
88. Matsumoto, H, Ohsawa, Y, Takahashi, S, Akiyama, T and Ishiguro, O (1996) An expert system for start-up optimization of combined cycle power plants under NO_X emission regulation and machine life management. IEEE Transactions on Energy Conversion 11(2):414–422
89. Melas, V B (2006) Functional Approach to Optimal Experimental Design. Springer, Berlin
90. Miettinen, K M (1999) Nonlinear Multiobjective Optimization. Kluwer, Boston, MA
91. Mitani, K and Nakayama, H (1994) Feed formulation for live stock using interactive multi-objective programming techniques. Agricultural Systems 10(2):99–108
92. Mockus, J (1989) Bayesian approach to global optimization. Kluwer, Boston, MA
93. Mockus, J (1994) Application of Bayesian approach to numerical methods of global and stochastic optimization. Journal of Global Optimization 4(4):347–365
94. Montgomery, D C (2005) Design and Analysis of Experiments, 6th edn. Wiley, New York
95. Myers, R H and Montgomery, D C (2002) Response Surface Methodology, 2nd edn. Wiley, New York
96. Nakayama, H (1989) An interactive support system for bond trading. In: Lockett, AG and Islei, G (eds) Improving Decision Making in Organizations. Springer, Berlin, pp. 325–333
97. Nakayama, H (1991) Satisficing trade-off method for problems with multiple linear fractional objectives and its applications. In: Lewandowski, A and Volkovich, V (eds) Multiobjective Problems of Mathematical Programming. Springer, Berlin, pp. 42–50
98. Nakayama, H (1992) Trade-off analysis using parametric optimization techniques. European Journal of Operational Research 60:87–98
99. Nakayama, H and Asada, T (2001) Support vector machines formulated as multiobjective linear programming. In: Proceedings of ICOTA2001, pp. 1171–1178
100. Nakayama, H and Sawaragi, Y (1984) Satisficing trade-off method for interactive multiobjective programming methods. In: Fandel, G, et al. (eds) Interactive Decision Analysis, Proceedings of an International Workshop on Interactive Decision Analysis and Interpretative Computer Intelligence. Springer, Berlin, pp. 113–122
101. Nakayama, H and Yun, Y B (2006) Generating support vector machines using multiobjective optimization and goal programming. In: Jin, Y (ed) Multi-Objective Machine Learning. Studies in Computational Intelligence, Springer, Berlin, vol. 16, pp. 173–198
102. Nakayama, H and Yun, Y B (2006) Support vector regression based on goal programming and multi-objective programming. In: IEEE World Congress on Computational Intelligence, Paper No. 1536 (in CD-ROM)
103. Nakayama, H, Nomura, J, Sawada, K and Nakajima, R (1986) An application of satisficing trade-off method to a blending problem of industrial materials. In: Fandel, G, et al. (eds) Large Scale Modelling and Interactive Decision Analysis. Springer, Berlin, pp. 303–313
104. Nakayama, H, Mitani, K and Yoshida, F (1992) A support system for multi-objective feed formulation. In: Proceedings of Software Conference, pp. 17–20
105. Nakayama, H, Kaneshige, K, Takemoto, S and Watada, Y (1995) An application of a multi-objective programming technique to construction accuracy control of cable-stayed bridges. European Journal of Operational Research 87(3):731–738
106. Nakayama, H, Arakawa, M and Sasaki, R (2002) Simulation-based optimization using computational intelligence. Optimization and Engineering 3:201–214
107. Nakayama, H, Arakawa, M and Washino, K (2003) Optimization for black-box objective functions. Optimization and Optimal Control 185–210

108. Nakayama, H, Yun, Y B, Asada, T and Yoon, M (2005) MOP/GP models for machine learning. European Journal of Operational Research 66(3):756–768
109. Nakayama, H, Inoue, K and Yoshimori, Y (2006) Approximate optimization using computational intelligence and its application to reinforcement of cable-stayed bridges. In: Zha, X F and Howlett, R J (eds) Integrated Intelligent Systems for Engineering Design. IOS, Amsterdam, pp. 289–304
110. Novikoff, A B (1962) Satisficing trade-off method for interactive multiobjective programming methods. In: Symposium on the Mathematical Theory of Automata, Polytechnic Institute of Brooklyn, pp. 615–622
111. Okabe, T (2004) Evolutionary Multi-Objective Optimization – On the Distribution of Offspring in Parameter and Fitness Space. Shaker, Aachen
112. Okabe, T, Jin, Y and Sendhoff, B (2003) Evolutionary multi-objective optimisation with a hybrid representation. In: Proceedings of Congress on Evolutionary Computation (CEC-2003), pp. 2262–2269
113. Okabe, T, Jin, Y and Sendhoff, B (2005) A new approach to dynamics analysis of genetic algorithms without selection. In: Proceedings of Congress on Evolutionary Computation (CEC-2005), pp. 374–381
114. Orr, M J (1996) Introduction to radial basis function networks. http://www.cns.ed.ac.uk/people/mark.html
115. Pareto, V (1906) Manuale di Economia Politica. Societa Editrice Libraria, Milano, Translated into English by A. S. Schwier, *Manual of Political Economy*. Macmillan, New York
116. Powell, M J D (1981) Approximation Theory and Methods. Cambridge University Press, Cambridge
117. Pukelsheim, F (2006) Optimal Design of Experiments. SIAM, Philadelphia, PA
118. Qian, Z, Yu, J and Zhou, J (1993) A genetic algorithm for solving mixed discrete optimization problems. Advances in Design Automation DE-Vol. 65-1:499–503
119. Quinlan, J R (1979) Discovering rules from large collections of examples: A case study. In: Michie, D (ed) Expert Systems in the Micro-Electronic Age. Edinburgh University Press, Edinburgh, pp. 168–201
120. Ragsdell, K M and Phillips, D T (1976) Optimal design of a class of welded structures using geometric programming. ASME Journal of Engineering for Industry 98:1021–1025
121. Rao, C R (1973) Linear Statistical Inference and Its Applications, 2nd edn. Wiley, New York
122. Rao, S S (1996) Engineering Optimization. Wiley-Interscience, New York
123. Rao, C R, Toutenburg, H, Shalabh and Heumann, C (2007) Linear Models and Generalizations: Least Squares and Alternatives, 3rd edn. Springer Series in Statistics, Springer, Berlin
124. Rivoirard, J (1978) Two key parameters when choosing the Kriging neighborhood. Mathematical Geology 19:851–856
125. Rockafellar, R T (1970) Convex Analysis. Princeton University Press, Princeton
126. Sacks, J, Welch, W J, Mitchell, T J and Wynn, H P (1989) Design and analysis of computer experiments (with discussion). Statistical Science 4(4):409–435
127. Sandgren, E (1990) Nonlinear integer and discrete programming in mechanical engineering systems. Journal of Mechanical Design 112:223–229
128. Santner, T J, Williams, B J and Notz, W I (2003) The Design and Analysis of Computer Experiments. Springer, Berlin
129. Sasajima, K, Inoue, K, Morishita, K, Hirai, J and Honda, M (2002) Study on the optimal anti-seismic designing method of cable-stayed bridge. In: Proceedings of the 3rd World Conference on Structural Control
130. Sawaragi, Y, Nakayama, H and Tanino, T (1985) Theory of Multiobjective Optimization. Academic, New York

References

131. Schaffer, J D (1985) Multiple objective optimization with vector evaluated genetic algorithms. In: Genetic Algorithms and Their Applications: Proceedings of the First International Conference on Genetic Algorithms, pp. 93–100
132. Schölkopf, B and Smola, A J (1998) New Support Vector Algorithms. Technical Report NC2-TR-1998-031, NeuroCOLT2 Technical Report Series
133. Schölkopf, B and Smola, A J (2002) Learning with Kernels: Support Vector Machines, Regularization, Optimization, and Beyond. MIT, Cambridge
134. Schonlau, M (1997) Computer Experiments and Global Optimization. Ph.D. Thesis, University of Waterloo
135. Schonlau, M, Welch, W J and Jones, D R (1997) Global Versus Local Search in Constrained Optimization of Computer Models. Technical Report, Institute for Improvement in Quality and Productivity, University of Waterloo, Canada
136. Seber, G A F and Wild, C J (2003) Nonlinear Regression. Wiley Series in Probability and Statistics, Wiley-Interscience, New York
137. Sengupta, D and Jammalamadaka, S R (2003) Linear Models: An Integrated Approach. World Scientific, Singapore
138. Shirakawa, M, Nakamoto, M and Hosaka, S (2005) Dynamic simulation and optimization of start-up processes in combined cycle power plants. JSME International Journal, Series B: Fluids and Thermal Engineering 48(1):122–128
139. Shirakawa, M, Kawai, K, Arakawa, M and Nakayama, H (2006) Intelligent start-up schedule optimization system for a thermal power plant. In: IFAC Symposium on Power Plants and Power Systems Control, Paper No. 549
140. Shirakawa, M, Arakawa, M and Nakayama, H (2007) Intelligent start-up schedule optimization system for a thermal power plant. Journal of Advanced Mechanical Design, Systems, and Manufacturing 1(5):690–705
141. Silverman, B W (1986) Density Estimation for Statistics and Data Analysis. Chapman and Hall, Boca Raton, FL
142. Simon, H (1957) Models of Man. Wiley, New York
143. Simpson, T W, Booker, A J, Ghosh, D, Giunta, A A, Koch, P N and Yang, R J (2004) Approximation methods in multidisciplinary analysis and optimization: A panel discussion. Structural and Multidisciplinary Optimization 27(5):302–313
144. Srinivas, N and Deb, K (1994) Multiobjective optimization using nondominated sorting in genetic algorithms. Evolutionary Computation 2(3):221–248
145. Stadler, W (ed) (1988) Multicriteria Optimization in Engineering and in the Sciences. Plenum, New York
146. Steinwart, I and Christmann, A (2008) Support Vector Machines. Springer, Berlin
147. Steuer, R (1986) Multiple Criteria Optimization: Theory, Computation, and Application. Wiley, New York
148. Stigler, G J (1950) The development of utility theory. Journal of Political Economy 58:373–396
149. Tamaki, H, Kita, H and Kobayashi, S (1996) Multi-objective optimization by genetic algorithms: A review. In: Proceedings of the 1996 International Conference on Evolutionary Computation (ICEC-1996), pp. 517–522
150. Törn, A and Žilinskas, A (1989) Global Optimization. Springer, Berlin
151. Tulkens, H (1993) On FDH efficiency: Some methodological issues and applications to retail banking, courts, and urban transit. Journal of Productivity Analysis 4:183–210
152. Ueno, N, Nakagawa, Y, Tokuyama, H, Nakayama, H and Tamura, H (1990) A multi-objective planning for string selection in steel manufacturing. Communications of the Operations Research Society of Japan 35(12):656–661
153. Vapnik, V N (1998) Statistical Learning Theory. Wiley, New York
154. von Neumann, J and Morgenstern, O (1943) Theory of Games and Economic Behavior. Princeton University Press, Princeton
155. Wang, G G and Shan, S (2007) Review of metamodeling techniques in support of engineering design optimization. ASME Transactions, Journal of Mechanical Design 129:370–380

156. Welch, W J, Buck, R J, Sacks, J, Wynn, H P, Mitchell, T J and Morris, M D (1992) Screening, predicting, and computer experiments. Technometrics 34(1):15–25
157. Wierzbicki, A P, Makowski, M and Wessels, J (2000) Model-Based Decision Support Methodology with Environmental Applications. Kluwer, Boston, MA
158. Yoon, M, Yun, Y B and Nakayama, H (2003) A role of total margin in support vector machines. In: Proceedings of International Joint Conference on Neural Networks, pp. 2049–2053
159. Yu, P L (1985) Multiple Criteria Decision Making: Concepts, Techniques and Extensions. Plenum, New York
160. Yun, Y B, Nakayama, H, Tanino, T and Arakawa, M (2001) Generation of efficient frontiers in multi-objective optimization problems by generalized data envelopment analysis. European Journal of Operational Research 129(3):586–595
161. Yun, Y B, Nakayama, H and Arakawa, M (2004) Multiple criteria decision making with generalized DEA and an aspiration level method. European Journal of Operational Research 158(1):697–706
162. Yun, Y B, Nakayama, H and Tanino, T (2004) A generalized model for data envelopment analysis. European Journal of Operational Research 157(1):87–105
163. Yun, Y B, Nakayama, H and Yoon, M (2007) Sequential approximate method in multi-objective optimization by using computational intelligence, 43(8):672–678. Transactions of the Society of Instrument and Control Engineers (In Japanese)
164. Zadeh, L A (1963) Optimality and non-scalar-valued performance criteria. IEEE Transactions Automatic Control AC8-1:59–60
165. Zeleny, M (1973) Compromise programming. In: Cochrane, J L and Zeleny, M (eds) Multiple Criteria Decision Making. University of South Carolina Press, Columbia
166. Žilinskas, A (1975) One-step Bayesian method of the search for extremum of an one-dimensional function. Cybernetics 1:139–144
167. Zitzler, E and Thiele, L (1998) An Evolutionary Algorithm for Multiobjective Optimization: The Strength Pareto Approach. Technical Report 43, Computer Engineering and Networks Laboratory (TIK), Department of Electrical Engineering, Swiss Federal Institute of Technology (ETH), Zurich CH-9092 Zürich, Switzerland
168. Zitzler, E and Thiele, L (1998) Multiobjective optimization using evolutionary algorithms – A comparative case study. In: Proceedings of Parallel Problem Solving from Nature V (PPSN-V), pp. 292–301
169. Zitzler, E and Thiele, L (1999) Multiobjective evolutionary algorithms: A comparative case study and the strength Pareto approach. IEEE Transactions on Evolutionary Computation 3(4):257–271
170. Zitzler, E, Laumanns, M and Thiele, L (2001) SPEA2: Improving the strength Pareto evolutionary algorithm. Technical Report 103, Computer Engineering and Networks Laboratory (TIK), Department of Electrical Engineering, Swiss Federal Institute of Technology (ETH)

Index

Symbols

$<_D$ 5
$\leq_D$ 5
$\leqq_D$ 4
μ-SVR 103
$\sim$ 4
$\succ$ 4
0-sensitivity 30

A

A-optimality 117, 118
acute 4
AIC 82
ALGDEA$_{fit}$ 160
α efficiency 64
alphabetical optimality 120
ant colony optimization 73
antiseismic property 169
approximate function 152
approximate optimal solution 152
approximate optimal value 152
approximate Pareto frontier 152
Arakawa 66
archive 53
archive truncation method 53
artificial intelligence 73
artificial neural networks 73
aspiration level 10
aspiration level approach 17, 22
augmented Tchebyshev scalarization
function 8, 23
automatic trade-off 25
auxiliary min–max problem 23
auxiliary scalar optimization 22
average fitness 48

B

back propagation method 74
Banker 57, 61
Bayesian approach 133
Bayesian global optimization 133
BCC model 56, 61
Bennett 89
best linear unbiased estimation 128
best linear unbiased
predictor 129, 130
bias 76
bias–variance trade-off 76
BIC 83
binary relation 4
Bishop 83
BLUP 129, 130
bond portfolio 34, 37

C

C-SVR 99
cable-stayed bridge 169
CART 73
CCR model 56, 60
CCR-efficiency 60
CCR-efficient frontier 60
cement production 34
center (of Gauss function) 80
Charnes 17, 56
compromise programming 21
computational intelligence 73
confidence ellipsoid 117
constant returns to scale 61
constraint qualification 12
constraint set 1
constraint transformation method ... 10
convergence 45
Cooper 17, 56

193

Index

correlation function 127
cost function 80
Cristianini 84
crossover 46
crossvalidation 83
crowded tournament selection 51, 53
crowding distance 51

D

D-optimality 117
DACE 114, 127
Dantzig 18
DASE 114
data envelopment analysis 56
De Jong 45
DEA 56
DEA efficiency 58, 60
DEA method 63
DEA-efficient 58
DEA-efficient frontier 58
DEA-inefficiency 58
Deb 48
decision maker 17
decision making unit (DMU) 56
density estimation 45
Deprins 58
design variable 46
DIDASS 22
Dinkelbach-type algorithm 40
discriminant function 73
distance-based criteria 120
diversity 45
DOE 114
dominated set 4
domination set 4
duality gap 20

E

E-optimality 117
Edgeworth 1
efficient frontier 3
efficient global optimization 126
efficient solution 2
EMO 151
empirical risk 76, 77
environmental selection 55
ε−constraint method 10
ε-insensitive loss function .. 98
erection management of cable-stayed
    bridges 34
Eschenauer 34
exact trade-off 29
expected prediction error 76
exterior deviation 86
external population 53

F

F-STOM 35
Farrell 56
FDH model 56, 61
feature space 84
feed formulation 34
Fiacco 12
fitness assignment in SPEA2 54
fitness evaluation 45
Fonseca 47
free disposable hull 58
Freed 74, 86
fuzzy logic 73
fuzzy mathematical programming 32

G

G-optimality 118
GA 45, 46
Gal 17
Gauss–Markov theorem 116
Gaussian function 80, 86
GDEA 64
GDEA efficiency 64
$GDEA_{fit}$ 66
GDEA method 64
generalization 76
generalization ability 84
generalized DEA 64
generalized least squares
    estimator 129
genetic algorithm 45, 46
geostatistics 126
Glover 74, 86
goal level 18
Goal programming 17
Goldberg 45, 47
GP 17
Grauer 22

H

Haimes 10
hard margin SVM 84
Holland 45

I

ID3 73
ideal value 23
incremental design 122
incremental learning 74
indifference relation 4
indifference set 4
induction rule 76
information matrix 116
interactive multiobjective
    programming 17

Index

195

interactive optimization methods ... 151
interactive programming method 17
interior deviation 84, 86
interpolation 79, 133

J

Jones 133

K

K-fold crossvalidation test 83
kernel function 85
Koopmans 2
Korhonen 30
Krige 126
Kriging 126
Kushner 133

L

Lagrange function 9
Lasdon 14
Latin hypercube 120
learning 76
least square estimator 78
least square method 77
leave-one-out method 83
likelihood 77
linear fractional min–max problem ... 40
linear independence 13
linear polynomial 123
linear predictor 130
linear programming 17
linear unbiased predictor 130
linearly weighted sum 7, 20
loss function 75
LSE 77, 78

M

machine learning 73
Mangasarian 74, 83
margin 84, 85
Matheron 127
mating selection 55
max–min distance optimality 121
maximum likelihood 76, 128
maximum likelihood estimator 128
mean squared error 76
membership function 32, 33
mesh-grid method 120
metamodel 114
Miettinen 17, 25
min–max distance optimality 121
MLE 128
MOGA 47
MONK's problem 96

MOP 1
Morgenstern 2
MSE 76
MSM 83
MSN 74
μ–ν–SVM 94
μ–ν–SVR 106
μ–SVM 93
multiobjective genetic algorithm 47
multiobjective programming
    problem 1
multisurface method 83
mutation 46

N

nadir value 23
Nakayama 15, 64, 66
natural genetics 45
natural selection 45
negative orthant 2
niche 48
niche count 48
NIMBUS 25
nondominated solution 2
nondominated sorting 51
nondominated sorting genetic
    algorithm 48
noninferior solution 2
normal equation 78
Novikoff 74
NSGA 48, 51
NSGA-II 51
ν_ε-SVR 108
ν-SVM 90
ν-SVR 101

O

offset 92, 94
offspring 51
operator P 22
operator T 22, 24
optimal design 115
Ordinary Kriging 132
overattainment 18

P

Pareto 1
Pareto frontier 3
Pareto optimal 2
Pareto solution 2
particle swarm optimization 73
pattern classification 73
payoff matrix 27
perturbation 26

plastic materials blending 34
population-based approach 46
positive orthant 2
power plants 176
predetermined model 110
prediction variance 116
preference order 4
preference relation 4
process variance 127
product correlation rule 127
production possibility set 58, 61
properly Pareto solution 24

Q

quadratic polynomial 123

R

radial basis function networks 74, 79
radial functions 79
radius (of Gauss function) 80
rank 47
ranking method 47
ratio efficiency 58
raw fitness 54
RBFN 74, 79
reference point method 17
regularization 78
relative efficiency 56, 58
response surface 114
returns to scale 57
Rhodes 56
ridge regression 78
Rockafellar 15

S

SAMO 152
SAO 113
satisficing trade-off method 22
Sawaragi 17
scalarization 5
scaled fitness 49
Schölkopf 84
Schaffer 46
Schonlau 133
second order optimality condition 13
selection 46
sensitivity analysis 29
sequential approximate multiobjective
    optimization 152
sequential approximate optimization   113
shared fitness 49
sharing function 49
Shawe 84
Simon 18

Simple Kriging 132
slack variable 88
Smola 84
smoothness parameter 127
soft margin SVM 88
space filling method 120
SPEA 53
SPEA2 53
Srinibas 48
startup scheduling 176
STEM 25
Steuer 17
Stigler 2
stochastic process 132
stochastic process model 127
strength Pareto evolutionary
    algorithm 53
strengths of dominators 54
strictly complementary slackness 13
support vector machine 74, 83
support vector regression 98
surplus variables 90
surrogate model 114
SVM 74, 83
SVR 98

T

Tanino 64, 66
Taylor 84
Tchebyshev scalarization function 8
technical efficiency 56
Thiele 53
total margin algorithm 91
tournament selection 55
trade-off 4
trade-off analysis 11
trade-off parameter 88
trade-off ratio 11
truncation method 55
Tulkens 58, 61

U

unbiased predictor 130
underattainment 18
Universal Kriging 132

V

value judgment 17
Vapnik 84
variance 76
VC dimension 84
vector evaluated genetic algorithm ... 46
vector inequality 2
VEGA 46, 47
von Neumann 2

Index

W

weak Pareto solution 3
weakly Pareto optimal 3
weight decay 80
weighted Tchebyshev scalarization
    function 8
Welch 133
width (of Gauss function) 80
Wierzbicki 17

Y

Yu 17
Yun 64, 66

Z

Zadeh 2
Zeleny 21
Zitzler 53